ON TIME TO THE DOCTORATE

A Study of the Increased Time to Complete Doctorates in Science and Engineering

Howard Tuckman
Susan Coyle
Yupin Bae

NATIONAL ACADEMY PRESS
Washington, D.C. 1990

NOTICE: The project that is the subject of this report was approved by the Governing Board of the National Research Council, whose members are drawn from the councils of the National Academy of Sciences, the National Academy of Engineering, and the Institute of Medicine. The survey project is part of the program of the Office of Scientific and Engineering Personnel (OSEP).

This report has been reviewed by a group of persons other than the author according to procedures approved by a Report Review Committee consisting of members of the National Academy of Sciences, the National Academy of Engineering, and the Institute of Medicine.

This report is based on research conducted by OSEP, with the support of the National Science Foundation, under NSF Contract No. PRA-8719855. Opinions, findings, conclusions, or recommendations expressed in this publication are those of the authors and do not necessarily reflect the view of the sponsoring agencies.

Library of Congress Catalog Card No. 89-62949
International Standard Book Number 0-309-04085-X

Additional copies of this report are available from : National Academy Press
2101 Constitution Avenue, N.W.
Washington, D.C. 20418

S010

Printed in the United States of America

PREFACE

A need exists for better models of what contributes to changes in the time that students take to complete the doctorate. Although time to the doctorate has been studied by Abedi and Benkin (1987), Berelson (1960), Prior (1962), and Wilson (1965), none of these studies are based on a causal model of student decisionmaking, and none consider the role of market forces in student decisions. The data presented in Chapter 1 suggest that time to the doctorate in science and engineering fields has been lengthening since 1967—in some fields, by as much as two years. Furthermore, it is anticipated that the lengthening trend will persist, at least into the near future, and have unfortunate consequences because of the decline in the college-age population and the dramatic increase expected in the number of job openings in the academic sector in the 1990s. In response, public policy makers are likely to become increasingly concerned with identifying and understanding ways to augment the supply of new doctorates. While shortages of this type are not expected for a few years, it is useful now to determine whether policies can be adopted that can limit or reverse the trend toward longer completion times in the science and engineering fields. Existing studies do not provide the information needed by policy makers to determine whether public policy could, or should, alter completion times sufficiently to slow or reverse the trends discussed in Chapter 1, or whether any policies can have a major impact on supply in the impacted fields.

The purposes of the present study are to render an in-depth analysis of what has happened to completion times since 1967, to provide a time-series data base for the period 1967-1986, and to develop a model that explains some of the factors that have caused an elongation to occur. This study looks at the effects of changes in five types of variables: family background characteristics, student attributes, financial aid, institutional environment, and market forces. Using data from the Doctorate Records File and the Survey of Doctorate Recipients maintained by the Office of Scientific and Engineering Personnel (OSEP) of the National Research Council and from other data sources,* the study develops a model to explain changes in both total time to the doctorate (TTD) and in the

* A more detailed description of the data from these sources is available on request from the National Research Council, Office of Scientific and Engineering Personnel.

several components of time to the doctorate. The model is then applied to 11 scientific and engineering fields: chemistry; physics and astronomy; earth, atmospheric, and marine sciences; mathematical sciences (including computer and information sciences); engineering; agricultural sciences; biological sciences; health sciences; psychology; economics; and all other social sciences.

This report is organized as follows. Chapter 1 begins with an examination of how and when time to the doctorate has been lengthening, illustrated by the rise in mean TTD from 1967 to 1986 in each of the 11 fields. Three components of TTD are introduced, and the mean values for each are presented and discussed. In addition, time coefficients allow one to contrast the way in which time to the doctorate has changed during the period, and two patterns of change are identified. Finally, quantitative estimates are provided of the person-year losses that society has incurred from the lengthening of completion time during this period. Chapter 2 reviews five avenues of inquiry in the literature as they relate to time to the doctorate and models of student decisionmaking. Chapter 3 introduces a causal model of the determinants of TTD based on an opportunity-cost framework of student decisionmaking. The role of financial aid and of market forces is explored in this context. Chapter 4 presents selected data on the zero-order correlations between the independent variables in the model and TTD (and its components). The correlations among the salary variables and unemployment/employment plans variables are discussed, and the contribution of each major vector (e.g., family background and student attributes) is examined. Chapter 5 introduces the statistical model and presents a summary of which regression coefficients are significant (and of their signs) for alternative specifications of the model. Several variants of the model are introduced to explore the effects of alternative measures of the key variables. Chapter 6 presents the regression coefficients for the basic model and several variants using registered time to the doctorate (RTD) as the dependent variable. Finally, Chapter 7 discusses the findings in this study, their implications, and research questions that warrant further study.

In addition, an extensive bibliography of readings on the determinants of student decisionmaking is provided (pp. 107-111). Appendix A (pp. 113-173) provides additional tables about (1) the components of TTD, (2) the person-year losses resulting from a lengthening of TTD, (3) variables in the model, (4) zero-order correlations among the independent variables, (5) several equations for estimating TTD, and (6) median total time to doctorate for the population as a whole and for selected demographic groups. Finally, acronyms used throughout this report are listed in Appendix B (pp. 175-177).

<u>Staff</u>

Howard Tuckman, consultant
Susan Coyle, staff officer
Yupin Bae, research associate
Linda S. Dix, editor

CONTENTS

LIST OF TABLES

LIST OF FIGURES

EXECUTIVE SUMMARY

This study provides an in-depth analysis of what has happened to doctorate completion times from 1967 to 1986, an aggregate time-series data base, and a model that explores some of the factors that cause an elongation of total time to the doctorate (TTD). The model looks at the effects of five types of variables: family background characteristics, student attributes, financial aid, institutional environment, and market forces. Using data from the Doctorate Records File and the Survey of Doctorate Recipients maintained by the Office of Scientific and Engineering Personnel of the National Research Council, a model is developed and tested to explain changes in TTD and in the several component parts of the TTD measure. The model is applied to 11 scientific and engineering fields: chemistry; physics and astronomy (P&A); earth, atmospheric, and marine sciences (EAM); mathematical sciences (including computer and information sciences); engineering; agricultural sciences; biological sciences; health sciences; psychology; economics; and all other social sciences.

Findings

Trends in TTD

The analysis finds that TTD, defined as the time lapse from the year that a student receives an undergraduate degree to the year that the doctorate is completed, initially decreased in the 1960s and then rose swiftly in the 1970s and 1980s. As a consequence, it now takes longer to complete a doctoral degree than at any previous time in this century. Mean TTD increased in each of the 11 fields in this study, ranging from a low of 0.3 years in economics to a high of 2.8 years in the health sciences. Increases in excess of two years were experienced in mathematics, psychology, and social sciences. Moreover, a double-digit percentage increase in TTD was experienced in all but biosciences and agricultural sciences. TTD increased even in fields where the time lapse to the doctorate was already quite long. For example, the average TTD in the health sciences was 10.5 years in 1967 and 13.3 years in 1986; in the social sciences it was 10.6 years in 1967 and 12.9 years in 1986. The evidence also suggests that student completion times are becoming more concentrated around the mean.

The rise in TTD is occurring at a nonlinear rather than a linear rate. In chemistry, physics and astronomy, and engineering, TTD has been rising at a decreasing rate. However, in the eight other fields examined, TTD has been rising at an increasing rate and is thus cause for greater concern.

Trends in Components of TTD

TTD can increase because students spend more time registered as students or because interruptions on the path from a bachelor's to a doctorate cause them not to be enrolled in school. Analysis of components of TTD indicates that most of the increase is attributable to the increase in registered time to degree (RTD)—that is, TTD less the time prior to graduate school entry (TPGE) and time not enrolled in graduate school (TNEU). In all of the 11 fields examined, RTD has increased substantially since 1967, accounting for most of the change in TTD in every case. Where RTD did not account for the total increase in TTD, interruptions in studies were the most frequent cause for lengthening of TTD. Delays in starting graduate school were an important additional explanation in only one field, health sciences.

Modeling TTD

Careful review of the relevant literature reveals five distinct but related lines of inquiry that bear on the development of a model of the causes of the rise in TTD. These lines of inquiry include the determinants of persistence and attrition, students' educational aspirations, the factors affecting enrollment in college, the role of expected returns and their effect on the decision to enter graduate school, and the literature on TTD. Several variables are consistently identified as affecting student choice: financial aid, whether the student is self-supporting, immediate background characteristics (rather than past background), quality of the undergraduate and graduate college, and differences in expected earnings and changes in market conditions.

The model used in the present study consists of five vectors of variables: family background characteristics, student attributes, tuition and financial aid, institutional environment and policies, and market forces. The model is estimated in both linear and nonlinear form and with two variants. Variant 1, the "common variables" model, includes the same variables for each field and is designed to determine whether a consistent set of variables is important in each field. Variant 2, the "unique variables" model, allows the number of variables in the explanatory equation to vary so that only those that are statistically significant are included in each final regression equation. For each field, regression equations are estimated using the 1967-1986 years as the units of analysis. Separate analyses made for the TTD and RTD variables produce the following results:

Results For TTD: Student characteristics and market forces are the key variables that affect TTD. However, the explanatory variables differ by field and by equation specification. The variable that most consistently explains rises in TTD is age at time of entry to graduate school. This is statistically significant in 9 of the 11 fields studied. Unfortunately, the model does not enable one to determine whether this variable relates to physical or intellectual effects of age (e.g., it takes older persons longer to learn) or whether its effects on TTD operate primarily because students who start later have a longer TPGE.

Among the market force variables, the salary ratio of doctorates 10 years after the doctorate to the salary of recent doctorates is significant in chemistry and EAM (using the common variables linear model) and in agricultural sciences and psychology (using the unique variables model). The salary level of doctorates 10 years after the degree is statistically significant in economics and social sciences. Among the family background variables, female gender is statistically significant in EAM and marine sciences. Type of institution attended affects TTD in some fields and quality of undergraduate institution (but not quality of graduate institution) is usually statistically significant. In psychology, a 1 percent increase in the percentage of a doctoral cohort with a bachelor's degree from a top 70 institution is associated with a 0.1 year decrease in TTD.

Results for RTD: No one variable is consistently large enough or consistently statistically significant enough across fields to explain the observed increase in RTD in all fields. Instead, different combinations of variables explain the rise in RTD in each of the 11 fields. In those equations where age is statistically significant, it tends to have a large impact on RTD. In the common variables log model, for example, the coefficients of the models range from 0.9 years (health sciences) to 6.4 years (social sciences). Since RTD is purged of TPGE, age does not act as a measure of late arrival at graduate school and, hence, its meaning is somewhat clearer in these regressions. Perhaps in part as a consequence, the age variable is not statistically significant in as many fields in the RTD equations (4) as it is in the TTD equations (9).

Financial aid that reduces student reliance on outside employment can make a difference in terms of RTD, and the type of aid is important in determining RTD as to which type of aid is most likely to reduce RTD, the models do not permit a single statement that applies to all fields. Instead, the effects of financial aid are highly field-specific. For example, a 1 percentage point change in federal support reduces RTD by 0.06 percent in EAM, 0.11 year in biological sciences, 0.23 in health sciences, and 0.09 in economics. Teaching assistantship (TA) support reduces RTD in EAM but increases it in biological sciences; and research assistantship (RA) support reduces RTD in math but raises it in biological sciences. The effects of particular forms of aid warrant further exploration.

In the fields of chemistry, mathematics, and economics, increases in the percentage of students with baccalaureates increase RTD in the common variables log model. Changes in market variables, particularly in the unemployment rate and the salary ratio, also affect RTD. Specifically, in the common variables log model, increases in the unemployment rate of 4-year college graduates tend to reduce RTD. A 1 percentage point change in the variable causes a 0.07 decline in TTD in EAM and a 0.02 decline in biological sciences. In the unique variables model, an increase in the percentage of new graduates seeking (but not yet finding) a position prior to graduation raises RTD in the biological sciences. Finally, increases in salaries for those who already hold doctorates, relative to increases in the salaries of new doctorates, have the effect of reducing RTD. This phenomenon is found primarily in the unique variables model and primarily in chemistry, mathematics, biological sciences, health sciences, psychology, and economics (Note: Several ratios are constructed with different years in the denominator, and which ratio is statistically significant is field specific).

Additional research on the sources of the rise in TTD is warranted. The process of acquiring a doctorate is a complex one that involves a variety of decisionmakers. No one set of unique factors adequately explains the rise in TTD and RTD. Moreover, our findings lack robustness with respect to the determinants of TTD and RTD. This may, in part, be attributable to lack of sufficient independent variation in the doctoral cohort's average annual time-series data for the period 1967-1986. For example, although time-series analysis did not indicate large and uniformly statistically significant effects for the student aid variables, simple cross-tabulations for 1986 and 1987 show that students reporting primary support from "own" earnings take, on average, over five more years to complete the doctorate than those with external financial aid. While this difference may be attributable to differences in the abilities and knowledge of recipients and non-recipients, we cannot rule out the possibility that a study of individuals would produce a stronger role for the financial variables. It may well be that alternative units of analysis will produce different and/or more consistent results than those presented here.

Conclusions

The data in this report indicate that students in general now take longer to complete their doctorates than at any previous time in this century. This exploratory analysis of the factors underlying these trends revealed a complex process that is affected by a variety of factors including availability of student support, labor-market conditions, sociodemographic characteristics of the degree recipients, and characteristics of both undergraduate and graduate degree-granting institutions. As noted earlier, no one of these factors consistently explained the pervasive upward trend that was found. Thus, more effort will be required to enhance understanding of this process.

Moreover, the authors did not explore the consequences of these trends, although the rising trend in TTD found in this study might lead to unacceptably high levels in some fields. First, increases in TTD lengthen the amount of time required for the supply to respond to any shifts in market demand. Such lags in supply responsiveness are costly to society. Second, increases in TTD may raise the costs and lower the returns to investment in doctoral training with possible consequences for career choice decisions of potential doctoral students. Other things equal, higher costs and lower returns can discourage students from pursuing training at the doctoral level. In addition, given the decision to pursue such training, increasing TTD may encourage some students to drop out before completing their degrees. Finally, lengthening TTD may, other things equal, reduce productivity by reducing the number of years spent by cohorts of newly produced degree-holders working as doctorates. Little is currently known about these possible consequences, but they are potentially serious enough to merit further attention.

1
WHAT HAS BEEN HAPPENING TO TIME TO THE DOCTORATE?

While factors leading to attainment of the doctoral degree have attracted research attention over the last 30 years, only recently has interest focused on the length of time it takes to earn the degree. Surprisingly, most current studies seem to overlook the phenomenon of increasing time to the doctorate occurring over the last two decades. Aggregate data on doctoral degrees show that while median time to the doctorate decreased in the 1960s, the decline was followed by a rather swift and steep increase through the 1970s and 1980s (Figure 1). Although lengthening degree time might simply reflect a distributional shift from doctorates in fields in which time to the doctorate is short (such as physical sciences and engineering) to those in which it is longer (such as humanities and education), other studies have found the increase is occurring in all fields (Coyle, 1987).

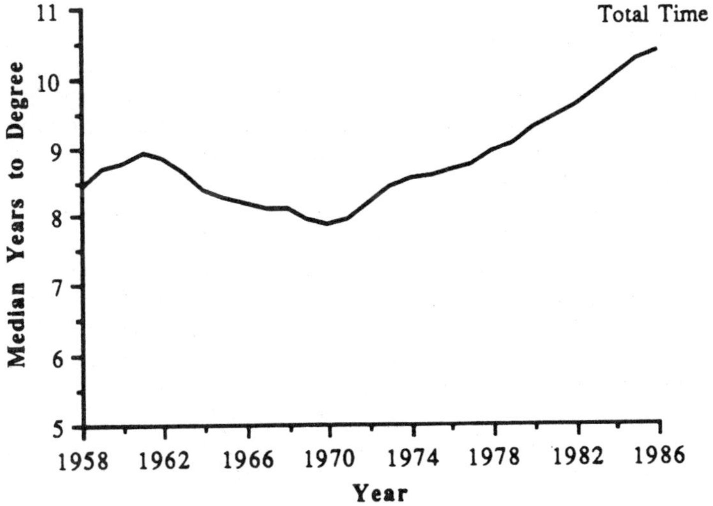

Figure 1 Median years to the doctorate, all fields combined including humanities and education fields, 1958-1986.

Components of Time to the Doctorate and How They Have Changed Through Time

The Several Kinds of Time

The time required to complete the doctorate can be measured in a number of ways, and the type of measurement used affects the degree of observed change as well as conclusions about which factors led to that change. The most comprehensive measure of time is total time to the doctorate (TTD), defined as the time from receipt of an undergraduate degree to completion of the doctorate. TTD is particularly useful for "pipeline" studies that examine the availability of new doctorates to enter the labor force. Similarly, TTD is useful for determining how quickly the supply of doctorate-level personnel will respond to changes in the demand for people with doctorates. Other things being equal, for example, a 10-year TTD would mean a delayed response of new doctorates to an increase in demand and a long wait for employers wanting to hire them.

Time to the doctorate also can be measured by the length of time that a student is actually registered in graduate school. Registered time to doctorate (RTD) is defined as TTD less the length of time prior to graduate entrance (TPGE) and any other time not enrolled in the university (TNEU)—that is, RTD = TTD - (TPGE + TNEU). TPGE may consist of service in the armed forces, time spent in travel, leisure or home-related activity, and/or postbaccalaureate work experience. There are two additional elements of RTD for which we have no measure: time spent in actual study/work toward the degree and time spent at the university in other pursuits. RTD is not a measure of the minimum time needed to complete the doctorate, since time spent in nondoctorate-related activity is also included. RTD, like TTD, is a measure of how quickly supply can respond to demand. In addition, it can be used as an indicator of the need for faculty and other resources in a graduate program. The relationship among these four time measures is summarized in Table 1.1.

Mean TTD for each of 11 science and engineering fields—chemistry; physics and astronomy ("P&A"); earth, atmospheric, and marine sciences ("EAM"); mathematical sciences, including computer and information sciences ("math"); engineering; agricultural sciences; biological sciences ("biosciences"); health sciences; psychology; economics; and all other social sciences ("social sciences")—is taken from the Doctorate Records File (DRF), the data base of the Survey of Earned Doctorates conducted annually by the National Academy of Sciences' Office of Scientific and Engineering Personnel (see, for example, Coyle, 1987: Table 2). Mean TTD, rather than median TTD, is used because it is more sensitive to small yearly changes in the data and easier to compare

TABLE 1.1: The Relationship Between the Several Time Measures

	Year of Undergraduate Degree Completion
+	Time Spent Prior to Graduate Entry (TPGE)
=	Year of Entrance into Graduate School
+	Time Spent at the University Working on Degree or Other Pursuits (RTD)
+	Time Spent Not Enrolled at the University (TNEU)
=	Year of Graduation with a Doctoral Degree (TTD)

among fields.[1] Although mean values can sometimes be distorted by the existence of a few outliers in the data, we did not encounter evidence of this problem (see Appendix Tables 2.1-2.5).

The time required to complete the doctorate has been increasing in the sciences and engineering primarily because students are spending more time in graduate school (i.e., RTD is rising). Figure 2 contrasts the growth of RTD with changes in its component measures, TPGE and TNEU.[2] The effects of changes in the intervening years are explored in the next section.

Mean Total Time to the Doctorate

Mean TTD increased in each of the 11 fields from a low of about four months in economics to a high of nearly three years in the health sciences (see Appendix Table 2.1). All but biosciences and agricultural sciences experienced double-digit percentage increases in TTD. The greatest increase, 30 percent, was in math, and TTD lengthened significantly even in fields in which it already was quite long. For each field, the within-year variation in TTD decreased from 1967 to 1986, suggesting student completion times more concentrated around the mean.

---

[1] Means are also used to provide the estimates of person-year losses shown on pp. 22 of this chapter. Although our analysis is confined to a discussion of mean times, median times have also been increasing (see Appendix Table 1).

[2] Appendix Tables 2.1-2.5 display the mean TTDs, TPGEs, RTDs, and TNEUs and their respective standard deviations, for each of the 11 fields at two points in time: 1967 and 1986.

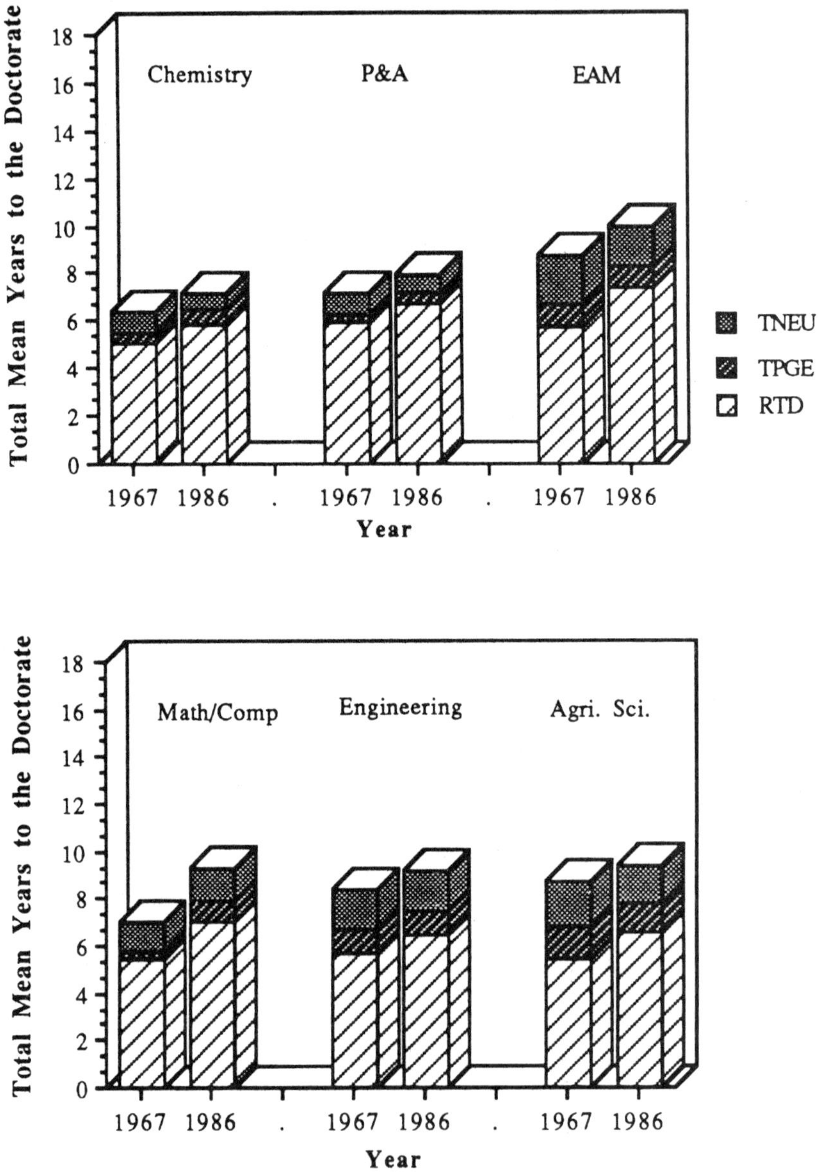

Figure 2 Components of mean total years to the doctorate, by field, 1967 and 1986.

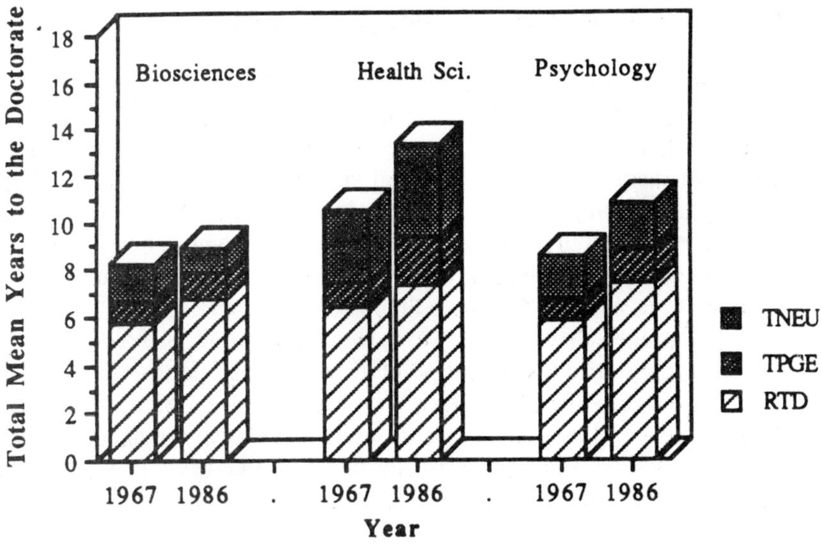

Total Mean Years to the Doctorate
18
16
14
12
10
8
6
4
2
0
Biosciences
Health Sci.
Psychology
1967 1986 . 1967 1986 . 1967 1986
Year
TNEU
TPGE
RTD

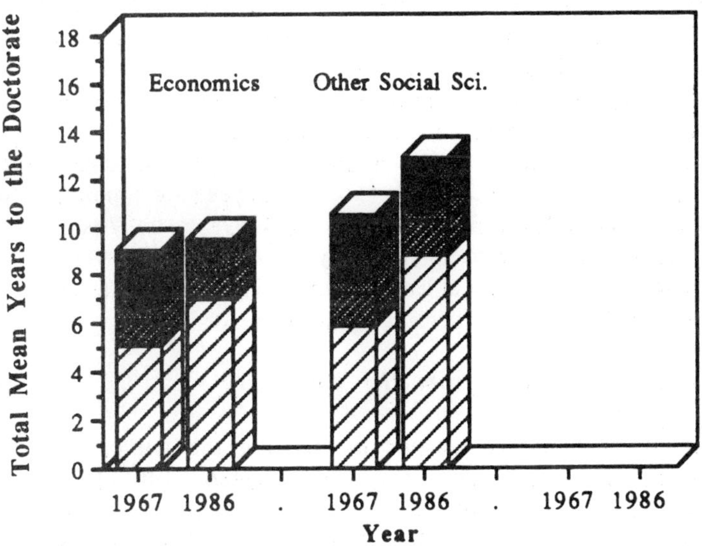

Total Mean Years to the Doctorate
18
16
14
12
10
8
6
4
2
0
Economics
Other Social Sci.
1967 1986 . 1967 1986 . 1967 1986
Year

Mean Time Spent Prior to Graduate Entrance

Changes in mean time spent prior to graduate school appear to have had little impact on the rise in TTD. TPGE showed little change. Except in health sciences, where there was an increase of approximately one year, on average, students in chemistry, P&A, EAM, and math entered graduate school less than one year after completing an undergraduate degree. Those in engineering, biosciences, agricultural sciences, psychology, economics, and social sciences spent between one and one-and-a-half years before entering graduate school. In health sciences, mean TPGE was a little over two years.

Although fairly large TPGE increases occurred in three fields—math, psychology, and health sciences—TPGE was a small portion of TTD in most fields. Two fields—EAM and agricultural sciences—experienced a decrease in TPGE. Analysis of the coefficients of variation for each year again revealed that within-year variance went down between 1967 and 1986, suggesting greater concentration of TPGE times around the mean.

Mean Registered Time to the Doctorate

Ideally, registered time to the doctorate should be broken down into time spent working toward the doctorate and time spent at the university in teaching or other activities unrelated to the doctorate (Berelson, 1960). Unfortunately, the DRF does not separately identify these two components. In all of the fields in the study, RTD increased at double-digit rates (see Appendix Table 2.3). Measured in both percentage and absolute terms, the largest increases occurred in the social sciences (where RTD rose from 5.9 to 8.8 years, or almost 50 percent) and economics (where RTD jumped from 5.1 to 7.0 years, or 37 percent). The smallest increases in RTD were in chemistry, P&A, and engineering. Overall, increases in RTD accounted for at least half of the increase in TTD and, in some fields, it accounted for over 100 percent of the increase.[3]

Mean Time Spent Away from the University

Students have many reasons for leaving the university prior to completing the doctorate. They may have financial difficulties, may be discouraged and/or frustrated with academe, or may need to seek additional data to finish the doctoral thesis (Dolph, 1983; Spady, 1970). Time not enrolled in the university increases TTD and, hence, is a variable worthy of separate

---

[3] This happened because decreases in the other components brought TTD down.

consideration. In most fields, TNEU decreased by at least half a year between 1967 and 1986. However, there was wide variability among the fields. For example, the decline was 1.5 years in economics, almost a year in health sciences, half a year in the biosciences, two-and-a-half months in math, and less than a month in psychology. Within-field variation for TNEU decreased in six fields and increased in five (see Appendix Table 2.4).

Summary

The major factor responsible for the change in TTD between 1967 and 1986 was the growth in RTD. In a majority of fields, a decline took place in TPGE and in TNEU (see Appendix Table 2.5).

The Nature and Significance of the Time Trend

The literature suggests it is now taking longer to complete a doctorate than at any other time. The upward slope of TTD follows a rather extended period of stability in time to the doctorate. In the near future, it will take even longer for doctoral candidates to complete their degrees.

The Two Models

The authors used statistical modeling to look at changes in TTD and other variables during each of the years between 1967 and 1986. For each field of study, regression equations were estimated using TTD or one of its three components as the dependent variable and time as the independent variable. Two different models, one which assumes that time has a linear effect [TTD = f(T)] and another which assumes a non-linear effect over time [TTD = f(T,T2)], were used. Using the linear model, for example, for chemistry students resulted in the conclusion that TTD increased by an average of 0.03 years per annum (or roughly 1 1/2 weeks per year) during the 1967-1986 time period (Table 1.2): a chemistry Ph.D. in 1967 took one-and-a-half weeks longer to complete the degree than in 1966 and nearly 30 weeks longer in 1985. Using the non-linear model, the increase in TTD for a chemistry doctoral candidate was about three weeks in 1966 and about 62 weeks in 1985.[4]

---

[4] These figures were determined as follows: the increase from 1966 to 1967 = 0.0632 years = (0.065 x 1) - (.0018 x 5) and from 1966 to 1985 = 1.20 years = (0.065 x 19) - (.0018 x 19).

The non-linear model produces a larger annual increase in TTD over time than the linear model for most fields and, with the exception of agricultural sciences, values derived from both the linear and non-linear models are statistically significant. In general, the non-linear model explained more of the variance than the linear model and, in most fields, provided a better fit of the data, hence a more accurate estimate of the effects of time on TTD.

Patterns of Change

In all 11 fields, there is a distinct and statistically significant upward trend in both TTD and RTD (Tables 1.2 and 1.3), although the trend is more pronounced for RTD than for TTD. For TTD, a non-linear time trend exists in most fields, suggesting that both the increase in time to the doctorate and the rate of change have differed across fields. Completion times accelerated in seven fields (EAM, agricultural sciences, biosciences, health sciences, psychology, economics, and social sciences) and accelerated and then decelerated in four (chemistry, P&A, math, and engineering).

For RTD, distinct patterns also emerge for each field, with some showing acceleration and others showing deceleration. A comparison of RTD and TTD suggests that in most fields the coefficients are quite close. This is not the case for the other components of time to the doctorate, however, suggesting that RTD is the factor most responsible for lengthening TTD.

An examination of time trend coefficients for the set of regressions using TPGE as the dependent variable shows that, in all fields, the amount of variation explained by time is less for TPGE than for RTD, in some cases half as much (Table 1.4). The non-linear model is preferable to the linear one in most fields, although in some fields its use has little impact on R^2. Using the non-linear model dramatically improves fit in the biosciences, economics, and social sciences; and it shows small gains in R^2 in math, engineering, health sciences, and psychology. The results again suggest that the time trend differs among fields.

The final set of regressions uses TNEU as the dependent variable (Table 1.5). In the linear model, mean time not enrolled in the university falls in seven fields (chemistry, P&A, EAM, agricultural sciences, biosciences, economics, and social sciences); rises in math, health sciences, and psychology; and remains stable in engineering. The results again suggest that the non-linear model provides better predictions of the time component in most of the fields.

TABLE 1.2: Estimated Time Trends in Mean Total Time to the Doctorate

	Time Coefficient Models						
	Linear Model			Non-Linear Model			
Field of Doctorate	Constant	T	R^2*	Constant	T	T^2	R^2*
Chemistry	6.482	0.03	0.46	6.344	0.065	-0.0018	0.46
		(3.91)			(5.30)	(1.93)	
Physics/Astronomy	7.273	0.04	0.60	6.887	0.14	-0.0050	0.78
		(5.19)			(33.17)	(17.43)	
Earth/Atmospheric/Marine	8.726	0.04	0.36	9.060	-0.054	0.0043	0.42
		(3.16)			(1.38)	(4.20)	
Math/Computer Sciences	6.832	0.12	0.90	6.978	0.076	0.0020	0.90
		(12.77)			(4.00)	(1.16)	
Engineering	8.353	0.05	0.60	8.046	0.13	-0.0040	0.68
		(5.25)			(15.8)	(6.84)	
Agricultural Sciences	8.731	0.005	0.01	9.036	-0.0078	0.0040	0.07
		(0.45)			(2.66)	(3.22)	
Biosciences	7.637	0.041	0.41	8.253	-0.13	0.0080	0.80
		(3.51)			(22.1)	(40.7)	
Health Sciences	9.616	0.13	0.58	11.105	-0.28	0.0190	0.92
		(5.03)			(36.05)	(82.32)	
Psychology	7.599	0.14	0.77	8.657	-0.15	0.0140	0.97
		(7.86)			(37.23)	(145.5)	
Economics	8.708	0.035	0.31	9.331	-0.13	0.0081	0.70
		(2.81)		(16.06)	(27.05)		
Social Sciences	9.626	0.12	0.68	10.844	-0.21	0.0160	0.97
		(6.23)			(80.01)	(215.9)	

*R^2 adjusted for the number of variables in the equation indicates the amount of variation in the dependent variable (TTD) that is explained by the time trend.

TABLE 1.3: Estimated Time Trends in Mean Registered Time to the Doctorate

	Time Coefficient Models						
	Linear Model			Non-Linear Model			
Field of Doctorate	Constant	T	R^2*	Constant	T	T^2	R^2*
Chemistry	5.169	0.03	0.72	5.024	0.072	-0.002	0.76
		(6.79)			(15.96)	(5.16)	
Physics/Astronomy	6.021	0.046	0.79	5.741	0.12	-0.004	0.92
		(8.36)			(75.65)	(31.38)	
Earth/Atmospheric/Marine	5.667	0.08	0.92	5.761	0.054	0.0012	0.91
		(13.96)			(5.12)	(1.23)	
Math/Computer Sciences	5.326	0.084	0.97	5.251	0.105	-0.001	0.97
		(22.67)			(47.18)	(1.92)	
Engineering	5.738	0.04	0.77	5.446	0.12	-0.0038	0.90
		(7.66)			(64.53)	(28.13)	
Agricultural Sciences	5.326	0.055	0.85	5.349	0.049	0.0003	0.84
		(10.17)			(4.36)	(0.075)	
Biosciences	5.549	0.055	0.91	5.762	-0.0029	0.0028	0.97
		(13.65)			(0.090)	(37.39)	
Health Sciences	5.792	0.064	0.74	6.255	-0.063	0.006	0.91
		(7.20)			(8.34)	(35.96)	
Psychology	5.452	0.090	0.86	5.954	-0.047	0.0065	0.97
		(10.49)			(9.72)	(89.31)	
Economics	5.228	0.099	0.94	5.101	0.13	-0.0016	0.94
		(12.12)			(32.22)	(2.28)	
Social Sciences	5.563	0.15	0.96	5.852	0.073	0.0038	0.97
		(20.85)			(8.79)	(10.79)	

*R^2adjusted for the number of variables in the equation indicates the amount of variation in the dependent variable (RTD) that is explained by the time trend.

TABLE 1.4: Estimated Time Trends in Mean Time Prior to Graduate Entrance

	Time Coefficient Models						
	Linear Model			Non-Linear Model			
Field of Doctorate	Constant	T	R^{2*}	Constant	T	T^2	R^{2*}
Chemistry	0.482	0.0045	0.21	0.494	0.0013	0.0002	0.13
		(2.23)			(0.02)	(0.15)	
Physics/Astronomy	0.360	0.005	0.49	0.362	0.0046	0.0000	0.43
		(4.14)			(0.75)	(0.0087)	
Earth/Atmospheric/Marine	0.867	0.0003	0.0004	0.921	-0.014	0.0007	0.00
		(0.082)			(0.73)	(0.81)	
Math/Computer Sciences	0.327	0.025	0.79	0.423	-0.0012	0.0013	0.83
		(8.33)			(0.011)	(5.76)	
Engineering	1.006	0.0007	0.003	0.924	0.023	-0.001	0.12
		(0.23)			(4.54)	(4.55)	
Agricultural Sciences	1.324	-0.014	0.42	1.365	-0.025	0.0005	0.37
		(3.64)			(2.35)	(0.49)	
Biosciences	0.685	0.012	0.38	0.882	-0.042	0.0026	0.83
		(3.35)			(29.54)	(51.84)	
Health Sciences	1.224	0.03	0.39	1.476	-0.039	0.003	0.47
		(3.43)			(1.38)	(4.52)	
Psychology	0.701	0.034	0.81	0.912	-0.024	0.0027	0.94
		(8.79)			(7.50)	(47.4)	
Economics	0.920	-0.0021	0.0065	1.196	-0.077	0.0036	0.43
		(0.34)			(16.21)	(16.24)	
Social Sciences	1.036	0.015	0.24	1.382	-0.080	0.0045	0.81
		(2.37)			(39.31)	(58.37)	

*R^2adjusted for the number of variables in the equation indicates the amount of variation in the dependent variable (TPGE) that is explained by the time trend.

TABLE 1.5: Estimated Time Trends in Mean Time Not Enrolled in University

	Time Coefficient Models						
	Linear Model			Non-Linear Model			
Field of Doctorate	Constant	T	R^2*	Constant	T	T^2	R^2*
Chemistry	0.807	-0.011	0.53	0.804	-0.01	-0.0000	0.47
		(4.49)			(0.92)	(0.006)	
Physics/Astronomy	0.879	-0.009	0.28	0.780	0.018	-0.0013	0.36
		(2.62)			(1.85)	(4.41)	
Earth/Atmospheric/Marine	2.103	-0.039	0.63	2.250	-0.080	0.0019	0.63
		(5.50)			(7.39)	(2.00)	
Math/Computer Sciences	1.113	0.011	0.21	1.160	-0.0018	0.0006	0.14
		(2.19)			(0.007)	(0.38)	
Engineering	1.627	-0.0003	0.0002	1.653	-0.0073	-0.0003	0.00
		(0.063)			(0.16)	(0.16)	
Agricultural Sciences	2.037	-0.036	0.52	2.299	-0.11	0.0034	0.60
		(4.43)			(12.68)	(5.95)	
Biosciences	1.345	-0.025	0.64	1.548	-0.081	0.0026	0.81
		(5.67)			(36.01)	(18.01)	
Health Sciences	2.473	0.041	0.27	3.345	-0.20	0.01	0.79
		(2.56)			(33.04)	(50.88)	
Psychology	1.364	0.015	0.17	1.765	-0.095	0.0052	0.72
		(1.94)			(28.23)	(39.85)	
Economics	2.394	-0.059	0.66	2.860	-0.186	0.006	0.83
		(5.95)			(41.56)	(20.54)	
Social Sciences	2.878	-0.400	0.42	3.488	-0.206	0.008	0.85
		(3.61)			(80.84)	(55.94)	

*R^2adjusted for the number of variables in the equation indicates the amount of variation in the dependent variable (TNEU) that is explained by the time trend.

The Shape of Change

The analysis shows time to the doctorate lengthened in all fields, and the time trend was non-linear for each of the three time components that make up TTD. Regression analysis revealed two distinct patterns in TTD (Tables 1.2 and 1.3). Data in eight fields was U-shaped, with a negative (or positive) T and a positive T^2 term, leading to an initial rise in TTD or a decline followed by an acceleration in TTD. This pattern existed in EAM, math, agricultural sciences, biosciences, health sciences, psychology, economics, and the social sciences. That TTD in these fields may continue to lengthen at an increasing rate over time is a source of potential concern. Data for three fields showed an inverted U shape, with a positive T and an negative T^2 term. For chemistry, P&A, and engineering, this pattern led to an eventual decline in the rate of increase in TTD over time. Figure 3 shows the actual data for each of the 11 scientific and engineering fields.

Since the non-linear time-trend model explained more of the variation in TTD than the linear model, it was used to forecast TTD for 1987. The results were then compared with the actual TTD values for 1987 (Table 1.6). The non-linear model closely projected TTD in 8 of 11 fields (within 0.01-0.34 year) but underestimated by close to half a year TTD in math/computer sciences, EAM, and agricultural sciences. The model produced a slight overestimate in TTD in the health and social sciences. For engineering, the projected and actual values were virtually the same.

Manpower Loss from Lengthening Total Time to the Doctorate

One important implication of the lengthening of TTD is that a given doctorate yields fewer potential person-years of labor force effort to society. The potential manpower loss calculated from increasing TTD does not equate to the total social implications of this trend. For example, no allowances are made for changes in the quality of new doctorates, market salaries, unemployment rates, on-the-job training times, or losses of Ph.D. positions at institutions that use predoctorates for research, teaching, or other work activities. Similarly, graduate students who might have been discouraged from obtaining a doctorate because of the time required to earn the degree are left out of the calculation. In addition, the baseline year used for the calculation is 1967. No presumptions are made as to whether the TTD in 1967 was better or worse than that which prevailed in some other year, since the goal is not to define the optimum year on which to

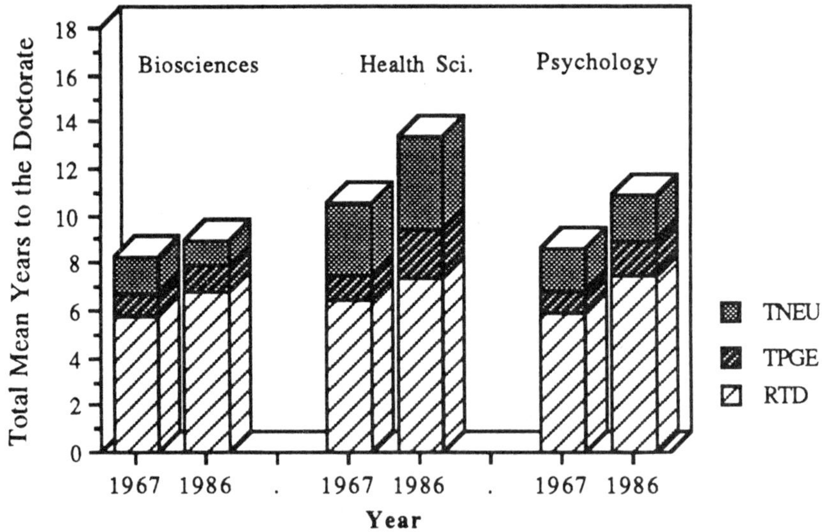

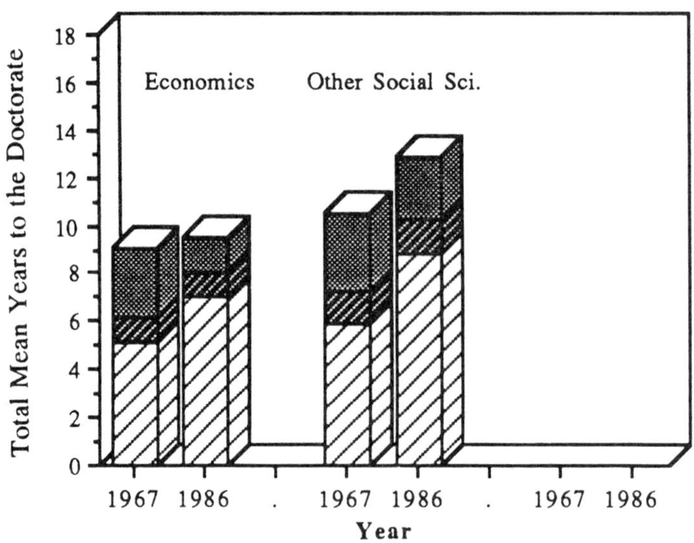

Figure 3 Mean total time to the doctorate, by field, 1967-1986.

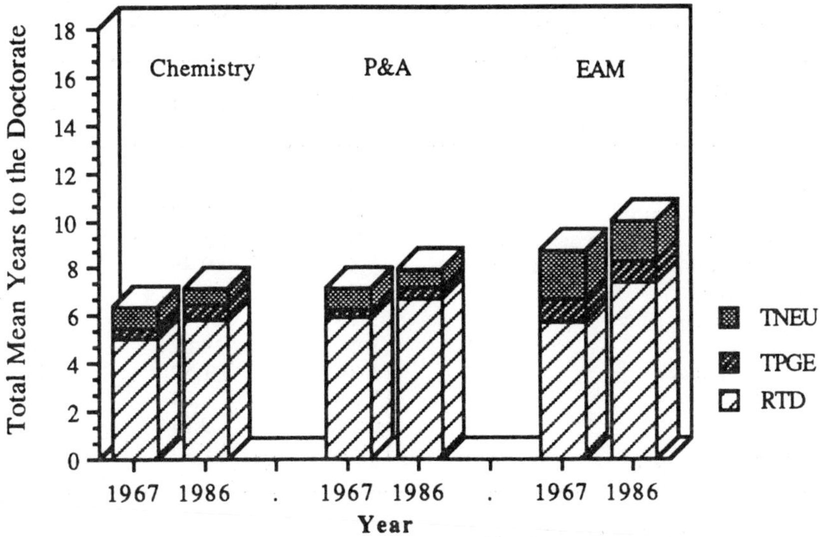

Total Mean Years to the Doctorate
18
16
14
12
10
8
6
4
2
0
Chemistry
P&A
EAM
TNEU
TPGE
RTD
1967 1986 . 1967 1986 . 1967 1986
Year

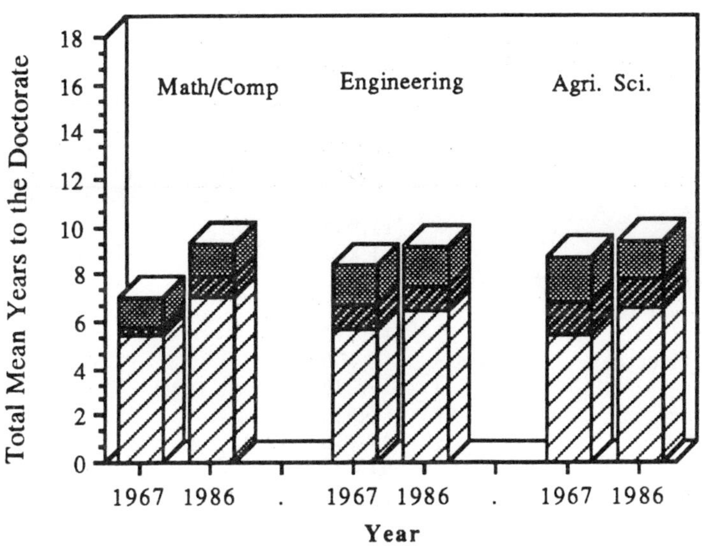

Total Mean Years to the Doctorate
18
16
14
12
10
8
6
4
2
0
Math/Comp
Engineering
Agri. Sci.
1967 1986 . 1967 1986 . 1967 1986
Year

TABLE 1.6: Difference Between Forecast and Actual TTDs, 1987

Field of Doctorate	Forecast*	Actual	Difference
Chemistry	6.90	7.05	-0.15
Physics/Astronomy	7.94	8.09	-0.15
Earth/Atmospheric/Marine Sciences	9.83	10.26	-0.43
Mathematics/Computer Sciences	9.41	9.86	-0.45
Engineering	9.04	9.05	-0.01
Agricultural Sciences	9.17	9.71	-0.54
Biosciences	9.11	9.02	0.09
Health Sciences	13.81	13.52	0.29
Psychology	11.58	11.24	0.34
Economics	10.07	9.74	0.33
Social Sciences	13.44	13.17	0.27

*Based on non-linear trend equation.

TABLE 1.7: Maximum Potential Person-Years Loss Resulting from Lengthening Total Time to the Doctorate, 1968-1986

Field of Doctorate	Estimated Number of Lost Person-Years	Loss as Percent of Total*
Chemistry	11,815	41
Physics/Astronomy	11,801	61
Earth/Atmospheric/Marine Sciences	3,872	40
Mathematics/Computer Sciences	13,306	85
Engineering	16,415	42
Agricultural Sciences	500	4
Biosciences	-17,082	-28
Health Sciences	5,529	63
Psychology	29,936	62
Economics	-1,885	-16
Social Sciences	8,751	27
Total	82,958	

*Determined by dividing "estimated number of lost person-years" by the total number of new doctorates provided during this period.

calculate TTD but, rather, to provide a quantitative estimate of how much high-level manpower has been lost over time.

To figure manpower loss, mean TTD calculated for each field and each year is subtracted from the 1967 mean TTD. The result is multiplied by the number of new doctorates in the given year to determine the manpower lost. The total loss for each field is calculated by summing the loss in each year beginning in 1968 and ending in 1986. A percentage loss is calculated by dividing the total person-years lost in all fields by the total number of new doctorates produced during this period. The calculation assumes all new doctorates are employed. Table 1.7 and Appendix Table 3 provide crude estimates of the potential gain in Ph.D. supply if TTD was reduced to the 1967 level. It should be noted, however, that these may be upper-limit estimates of the loss because many individuals pursuing the doctorate over an extended time simultaneously performed other work whose value to society cannot be determined. These figures do not take into account the effects of increases in TTD in discouraging career choice. Table 1.7 suggests that a small but meaningful increase in supply—greatest in psychology—could be achieved if the trend toward a longer TTD could be reversed.

2
MODELS OF THE FACTORS THAT AFFECT STUDENT CHOICE AND TIME TO THE DOCTORATE: A LITERATURE SURVEY

Over the past 30 years, an extensive literature has developed addressing TTD and the factors affecting student decisions to pursue postgraduate education. The literature has focused on five lines of inquiry: (1) persistence and attrition, or factors that cause students either to complete their education or to terminate it before a degree is received; (2) educational aspirations, or students' plans for pursuing additional education and training; (3) enrollment in college, which is similar in focus to the literature on aspirations but often uses different assumptions and statistical approaches to study the problem of student choice; (4) expected or perceived value of investing in education; and (5) TTD. This review is selective in nature, focusing mainly on findings that aid in an understanding of student choice.

Literature on Persistence and Attrition

The focus of much of the early research on attrition identified factors that caused students to quit school at the undergraduate and graduate levels (Berelson, 1960; Summerskill, 1962), not the processes that caused individuals to drop out or the quantitative impact of the factors involved. Descriptive approaches of this type can still be found in the current literature (e.g., Teague-Rice, 1981; Dolph, 1983), but more recent studies, beginning with the work of Spady (1970) and the model proposed by Tinto (1975), focus on causality.

The model by Vincent Tinto (1975) is important because it explains how the interaction of many factors affects decisions to remain in school or to drop out. Longitudinal and theoretical in nature, the model assumes individuals enter institutions with specific attributes, background characteristics, prior experiences, and commitments that are integrated into their academic and social lives. The institution itself may have important effects on grade performance, intellectual development, peer group interaction, and faculty interaction with students.

In Tinto's model, grade performance and intellectual development contribute to academic integration and thus to goal commitment, and peer-group and faculty-student interactions contribute to social integration and to the student's commitment to the institution. And the interplay between the individual's commitment to completing college and his/her commitment to the institution affects the decision whether to drop out and for how long.

Other researchers have used the Tinto model as a basis for regression and path analyses, and their findings tend to support Tinto's theory. For example, Pascarella and Terenzini (1983) found that social and academic integration have about equal effects on persistence and that students who are better integrated into an institution are more likely to complete the undergraduate degree than those who are not. Several other causal models are discussed in Bean (1980). For the most part, the factors identified by Tinto as having an impact on students are the same across studies, although some researchers differ as to which factors have direct and which have indirect effects. For example, Smart and Pascarella (1986) argue that schooling plays a direct role in determining social mobility. Differential levels of educational attainment yield different levels of achievement among persons with equivalent social backgrounds. Education also indirectly affects social mobility by serving as a "mediator" through which individual resources such as ability and background are converted into earnings and occupational status.

Two aspects of the Tinto model warrant further comment. First, the role assigned to quality of college is ambiguous in the theory. Some researchers have found that better colleges produce a "higher yield" of graduates from the entering class (Knapp and Goodrich, 1952; Knapp and Greenbaum, 1953) while others have suggested the opposite (Davis, 1966).[5] Many studies have looked at the role of college characteristics and college environment in affecting persistence and educational aspirations (see, for instance, Pascarella, Terenzini, and Hibel, 1978). Recent research shows that student interaction with faculty has a very small, albeit positive, effect on academic performance. In the Tinto model, faculty-student interaction affects persistence directly through its effect on social interaction and indirectly through its effect on grades. These two, in turn, affect

---

[5] Specifically, Tinto questioned whether students at higher-quality schools have lower expectations. Davis had posited a "frog pond" effect, wherein the higher the average ability of the student body, the lower the grades of individuals of given ability, as compared to the grades they would have received at institutions populated by students of lower ability. Since grades affect expectations and expectations affect dropouts, a person of given ability level may be more likely to drop out at a higher-quality than at a lower-quality institution.

academic integration and the decision to drop out (see Pascarella and Terenzini, 1979).

Second, the Tinto model relegates changes in economic factors, such as unemployment and expected future earnings, to a category called "external impacts upon dropout." Changes in these variables are assumed to affect persistence indirectly by operating on student commitment to finish school and to the educational institution itself but do not directly enter the model as observables. Interestingly, Tinto assumes that an individual goes through a benefit-cost calculation to determine if it is worthwhile to stay in college, but he ignores the role that opportunity-cost considerations might play. Some who have relied on the Tinto model have considered economic factors as of secondary importance, although most studies assign a role to financial aid (e.g., Ethington and Smart, 1986). The relegation of economic factors to a secondary role makes it difficult to study the impact of economic factors other than availability of funds on the decision to drop out and also precludes researchers from using the Tinto model to explore the effects of market forces on student choice.

Both descriptive and causal studies point to parents' education, student grade-point average (GPA), race, and educational characteristics as affecting student persistence at the undergraduate level. These studies also tend to validate the importance of the interaction between students and faculty in keeping students in school. Other student variables associated with high attrition rates are upbringing in a rural area, father with less than a high school education, religion, and separation from one's spouse.

Attrition at the doctoral level has been less carefully studied, is less well understood, and is most often expressed in descriptive rather than model form. For example, Tucker, Gottlieb, and Pease (1964) present data based primarily on student responses to questionnaires, indicating that the largest single reason for dropouts is student finances. Students without money to meet expenses or not having a teaching assistantship, research assistantship, or other financial aid were more likely to drop out than those with adequate financial support. Teague-Rice's (1981) study of female doctorates at Auburn from September 1971 to 1977 and Dolph's (1983) study of Georgia State students from 1970 to 1980 confirm the importance of scholarship, assistantship, or fellowship support. Students who are full-time, have a positive relationship with their dissertation chairperson, and score high on comprehensive exams also tended to remain on the doctoral track, according to these studies.

A recent causal analysis by Girves and Wemmerus (1988) used the Tinto model to explore "degree progress" at the graduate level. For doctoral-level students, academic involvement appeared to have a direct impact on degree progress, while for master's-level students, such involvement appeared not to be important. Moreover, social integration did not seem to plan an important role in students' persistence, suggesting Tinto's conceptualization may not be entirely valid at the graduate level. Grades were important determinants of persistence for master's-level students, while the effect of grades on degree progress disappeared at the doctoral level.[6] Girves and Wemmerus argue that involvement in the academic program, the role of the advisor, the number of faculty members a student gets to know, the faculty/student relationship, and the type of financial support are all important in affecting degree progress, but the effect of some of these variables is indirect. Differences among fields, identified by Biglan (1973), also are important in influencing a student's decision to complete a doctoral program, Girves and Wemmerus say.

Literature on Educational Aspirations

There is a substantial body of research that attempts to identify the reasons students decide to attend graduate school (e.g., Baird, 1976; Gropper and Fitzpatrick, 1959). More recent studies are less interested in "why" than in the process by which key variables interact to shape student educational aspirations. One major line of inquiry looks at how the structural and environmental characteristics of colleges influence students to seek graduate training [see, for example, Astin and Panos (1968)].

Pascarella's 1984 study, which used a causal model of educational aspirations based on Tinto's dropout model, finds that the direct effects of any single aspect of the college environment are "quite modest" and the best predictor of educational aspirations at the end of the second year of college is the level of educational aspirations at entrance to college. The only other factors directly affecting the decision to continue to a higher level of training are a student's cumulative GPA and a cumulative measure of college environment, according to Pascarella.

Other studies use somewhat different causal models and include different variables but, nonetheless, reach similar conclusions. Alwin (1974), for example, found that a small amount of the variation in student aspirations can be attributed to differences in the college environment after student inputs are controlled, and Heyns (1974) found that verbal achievement and curriculum placement affect the relationship between student inputs and student aspirations.

More recently, Ethington and Smart (1986) modified the Tinto model to test how the decision to enter graduate school is made. The model assumes the

---

[6] The variability in grades is probably small at the doctoral level. Thus, the finding of no effect does not necessarily imply that academic performance doesn't matter.

decision is the culmination of a series of choices made as students progress through the educational system, and it differs from those used in earlier studies by giving less influence to certain variables (for example, factors that affect decisions early in the choice process exert subsequent influences only indirectly). The Ethington and Smart model assumes that decisions regarding graduate education are based on "blocks" of independent variables that interact with each other. Student background characteristics and high school experiences comprise one block, which affects the choice of undergraduate institution. A second block measures student social and academic integration within the undergraduate institution, which in turn is influenced by the background "block" of variables. At a certain point, the effects of background characteristics wane as undergraduate experiences, financial aid, and receipt of the undergraduate degree replace them in importance. Enrollment in graduate school is dependent on all the measured variables, but results of the study indicate degree completion and receipt of financial aid have, by far, the greatest impact on graduate school enrollment. Student background characteristics have, at best, a marginal impact on the decision; the only student background variable showing a direct effect is the educational level of the student's family. For men, selectivity of the undergraduate institution has a strong positive effect on graduate school attendance while, for women, size of the undergraduate institution is important. Ethington and Smart found students with greater social and academic involvement in their undergraduate institutions are more likely to go to graduate school than those less involved.

Spaeth's (1968) study of factors that "allocate" college graduates to graduate and professional school, more empirical than theoretical, assumed that parental socioeconomic status (SES), students' intellectual ability, undergraduate academic performance, and the quality of the college from which they graduated influenced choice of graduate school. The Spaeth study looked at the career plans of 1961 college graduates, using a path-analytic model to relate quality of graduate school attended to student input and family background characteristics. Student undergraduate grades and the "intellectual caliber" of the undergraduate college attended were found to be major determinants of the quality of graduate school attended.

Literature on Enrollments

Some studies have equated enrollments with demand (Heath and Tuckman, 1986), although the former variable includes elements of supply while the latter does not. For example, Campbell and Siegal (1967) looked at demand for higher education using time-series data to estimate the ratio of undergraduate degree enrollments to the number of those eligible to enter undergraduate institutions. Likewise, Carroll *et al.* (1977) analyzed the effects of Basic Educational Opportunity Grants on enrollment decisions, and Alexander and Frey (1984) attempted to identify the determinants of enrollment in MBA programs. Most studies of these types focused on the direct effects of a set of independent variables and used regression analysis to identify what factors determine enrollment. But such studies largely ignore the interactive relationships captured by a path analysis, and many relegate sociological and psychological variables to a secondary role, although controlling for student characteristics such as race, age, and ability.

Researchers usually place great importance on the role of tuition, family income, and financial aid in determining enrollment. Some also assume that external economic variables, such as unemployment, affect enrollments. A number of literature reviews have explored factors that determine demand for higher education (Becker, 1986; Jackson and Weathersby, 1975; Leslie and Brinkman, 1986).

Heath and Tuckman's (1989) review found that early studies that relied on a net tuition variable (gross tuition less financial aid) were flawed because net tuition fails to recognize that changes in tuition and financial aid have different effects on student demand. It also found that type of financial aid was important, with the evidence suggesting fellowships have a larger effect on demand than teaching assistantships. The review also revealed that the price elasticity of demand is lower at high-quality undergraduate institutions than at other 4-year schools and is less for graduate education than for undergraduate education. Finally, the review showed that decreases in financial aid at the graduate and professional levels reduce matriculation, increase the dropout rate, and lengthen time to the doctorate.

Heath and Tuckman developed a model of the determinants of the demand for higher education that breaks the group of potential graduate students into five subpopulations: recent college graduates; persons in the work force; homeworkers who might return to graduate school; those discharged from the armed forces interested in higher education; and non-residents of the United States who attend U.S. institutions of higher education. Demand for graduate training in any given field is based on family characteristics, individual abilities and interests, tuition and financial aid variables, the characteristics of the educational organizations, and economic and social variables. The model can be used to explain the demand for both graduate and undergraduate training and, by introducing time notation, can also be used to explain persistence.

Literature on Expected Returns

Since the early 1960s, researchers have recognized and written about the economic returns from investment in schooling. Implicit is the assumption that fields with higher returns attract more students than fields with less lucrative returns. Some studies have shown that when salaries rise in a field (e.g., business), more students major in it (Berryman, 1982), while others have actually formulated tests designed to show a specific causal relation. For example, Koch (1972) computed internal rates of return by academic field and compared them to changes in enrollment in 17 major fields at Illinois State University. He found that a small group of students do indeed shift to fields where salaries are high.

A more recent and complete study by Cebula and Lopes (1982) looked at enrollment data for 28 fields at Illinois State University from 1973 through 1976 and confirmed that future earnings are an important consideration in selecting a major. But changes in earnings differentials were more important than the absolute value of the earnings differential, and neither the outlook for a given field nor Graduate Records Examination (GRE) scores were statistically significant predictors of field choice.

Freiden and Staaf (1973) introduced an opportunity-cost approach to the student-choice literature, albeit indirectly, arguing that students switch curriculum groups as they progress through college and acquire information about alternative educational opportunities. According to this approach, students prefer "bundles" of courses that fulfill specific degree requirements and tend to pursue curriculum groups in which they have a comparative advantage, as defined by their verbal and quantitative Scholastic Aptitude Test (SAT) scores.

More rigorous modeling of the relationship between enrollments and earnings potential in an academic field can be found in the work of Freeman (1971), who argued that differences in relative earnings signal potential students to enter fields experiencing shortages. He formulated a set of equations based on interactions between changes in starting salaries, government research and development expenditures, and student enrollments. Freeman showed markets adjust to changes in demand gradually and the nature of this time lag varies among fields. In some fields, a cycle of periodic shortages and excesses develops, emulating the cobweb pattern found in agricultural employment. Freeman's model has been tested and modified in the last decade, and while the cobweb pattern is in dispute, most research supports the conclusion that expected earnings affect student decisions (e.g., Hansen *et al.*, 1980).

Trusheim and Crouse (1981) examined the effect of relative earnings on student decisionmaking in a different way, focusing on the effects of college prestige and selectivity on income. They found that, for men, type of occupation depends heavily on having gone to college but not very much on the prestige or selectivity of the college attended. College selectivity did have a statistically significant effect on income, however.

In an interesting and provocative piece, Berger (1988) tested Freeman's assumption of student myopia by estimating conditional logit models that incorporate alternative predicted future earnings measures. Using data from the National Longitudinal Survey of Young Men for five broad fields of study, Berger used alternative earnings measures to see if students were more likely to choose a field of study based on its potential future earnings than on its starting salary. After controlling for background characteristics, he found the probability of choosing one field over another increases as the present value of its predicted earnings stream increases relative to that in other fields.

Researchers agree that relative income is important, but they disagree about how it should be measured. Should starting salaries, mid-career salaries, or future earnings profiles be used as a proxy for expected future earnings? Should salaries be measured relative to a numeraire field (e.g., a common base) or in absolute terms? Should the salary average be for a field or an occupation? These and related questions are addressed in future chapters.

Literature on TTD

Literature on the factors determining TTD is limited. Interest in TTD emerged in the early 1960s, when demand for graduate education led to a temporary shortage of Ph.D.s. Early studies by Berelson (1960) and Carmichael (1961) used survey analysis and data provided by the National Research Council to explore what was happening to TTD over time. Among Berelson's findings were that TTD can be shortened if full-time support is provided to a large number of doctoral students. Shortening TTD will allow more students to be educated, Berelson found, but it would do more to increase the quality of training than to increase the number of available places. He also found that the main cause of the rise in TTD was time spent in nondoctorate-related pursuits, such as work as a teaching assistant or research assistant, or time spent in work-related pursuits. Berelson's work contains little information on the background characteristics of students and how they have changed through time. Although it does not address the interactions between students and their environment in model form, it does suggest specific institutional policies that might shorten TTD.

Early on, researchers realized that "the Ph.D. is an open-end degree [that] cannot be circumscribed by an exact preordained time limit" (Prior, 1962). Prior's work, like that of Berelson and Carmichael, provides useful information on institutional policies, but it does not explain changes in TTD nor does it show the quantitative effects of the various factors causing increased TTD.

A study by Wilson (1965), based on a questionnaire sent to graduates, graduate deans, and departmental representatives in a representative group of fields at 23 doctoral institutions, is more useful, since it identifies the factors that affect TTD. Graduate deans, graduate faculty, and doctorate recipients all felt discontinuity of attendance, work as a teaching assistant, and writing the dissertation off-campus contributed to increased TTD. Similarly, financial problems, inadequate preparation in a foreign language, lack of coordination between beginning and advanced stages of graduate work, family obligations, inadequate undergraduate preparation in the major, and transfers among graduate institutions were named by all three groups as factors leading to lengthened TTD, the study found. Continuity of study and adequate time to devote to study were seen as key to rapid completion of the doctorate. Clarity of institutional and departmental expectations regarding doctoral requirements were cited by deans as critical. Respondents to the Wilson questionnaire made two recommendations of special note: (1) students need to be insured adequate amounts and appropriate forms of financial support so they minimize their reliance on nondoctorate-related employment and (2) expectations of the skills and competencies that doctoral candidates have should be better articulated.

While the Wilson study is thorough and thought-provoking, it does not provide insight into the role of student input variables in TTD, nor does it provide a quantitative estimate of institutional impacts. Abedi and Benkin (1987) attempted to fill this gap by studying over 4,000 students who received doctoral degrees from UCLA between 1976 and 1985. The Abedi-Benkin study postulated two regression equations with mean TTD as one dependent variable and mean RTD as the other. Three key sets of independent variables—demographic, financial, and academic—were included in the analysis.

Using stepwise regression to find the statistically significant variables, the authors found that source of support was the most important predictor of TTD (using the F-ratio as the criterion for importance), while "postdoctoral plans" was the second most important. Average TTD was lower for those in the postdoctoral study/trainee category than for those who planned to enter the labor force after receiving their degree, suggesting many who plan to enter the labor force post-degree are already employed, perhaps slowing their progress toward the doctorate. Other significant variables were the number of dependents, sex, and field of study.

Summary

A great deal of research has been conducted on what determines student decisions regarding higher education. There is more literature assessing decisionmaking at the undergraduate rather than the graduate level. Similarly, more studies have assessed decisionmaking in the fields of education and sociology than in economics. Studies have moved in the direction of causal modeling and away from pure empirical analysis. The literature review suggests that many recent studies performed by non-economists have relegated economic factors to a secondary role and the significance of market forces in student decisionmaking has been neglected.

There is a dearth of studies about time to the doctorate—partly because researchers seem to have lost interest in the question when the shortages of the early 1960s turned to surpluses in the 1970s, and partly because researchers seem unaware of the trend toward increasing TTD.

Most studies of aspirations, dropouts, enrollments, and expected returns were noncausal and largely descriptive in the early 1960s, giving way in the 1970s to more formal modeling (path analyses or deterministic demand models). The Tinto model provided the basis for much subsequent educational and sociological research, but it failed to integrate the economic variables considered important in studies of enrollment and expected returns. And most studies in these latter two areas have tended to ignore demographic and sociological variables, while others have not paid adequate attention to institutional environment.

Overall, findings from several avenues of inquiry have not been integrated into comprehensive theory of what determines time to the doctorate and, as a result, studies of TTD have been largely noncausal and empirical. Despite this, several variables appear to affect student choice consistently:

- Financial aid (this raises the question of whether the variable is also important in determining TTD);
- Main source of support (the literature provides little insight into the quantitative importance of this variable in determining TTD);
- Immediate, rather than past, background characteristics (for example, current grades are more likely than past ones to affect current decisions. Many socioeconomic factors that affect the decision to enter college—for example, parent's education and income—are unlikely to have a major effect on TTD at the doctoral level. Work is needed on personal factors that have an immediate effect on TTD);
- Quality of the undergraduate and graduate college (at present, little is known on the quantitative effects of organizational environment on TTD); and
- Differences in expected earnings and changes in market conditions (to date, such variables have not been added into models of TTD).

These insights are the basis of the theory and model discussed in the following chapters.

3
A MODEL OF THE DETERMINANTS OF TTD

The theory developed in this chapter is based largely on an opportunity-cost explanation of student choice and is concerned primarily with direct, rather than indirect, effects. The chapter begins with a discussion of five vectors that belong in a model of student choice and then explores the opportunity-cost arguments underlying the choice of variables. The ways in which financial aid affects time to completion and the role of market forces (relative salaries and employment opportunities) also are examined. A discussion of the variables used in the model is included.

The Model

TTD is directly affected by five vectors of variables similar to those shown to influence demand for graduate school and persistence to the degree: family background characteristics (F), individual abilities and interests (I), tuition and financial aid (TLFA), environment and policies of institutional organizations (O), and economic and social forces (E).[7] The relation is:

$$(1)\quad TTD_{dt} = f_d(F_{d(t\text{-}n)}, I_{d(t\text{-}n)}, TLFA_{d(t\text{-}n)}, O_{d(t\text{-}n)}, E_{d(t\text{-}n)})$$

In the formula, "dt" denotes the field (d) and the year (t) in which a given cohort of doctorates received Ph.D.s. Since the model is used to explain changes in TTD in 11 fields and many of its variables affect TTD several years before a cohort receives the doctorate, a cohort that received its doctorate in year "t" is assumed to have been affected by the variables in the five vectors "n" years prior to the time that the degree was completed (although not acknowledged in Equation 1, the "n" may be different for each variable and also for each field).

---

7 The model can be formulated in path-analytic terms by having F affect O, I, and E and by considering the indirect effects of E on O, TLFA, and F.

The doctorate cohort (i.e., those receiving their Ph.D.s in a given year) is used as the unit of analysis, although the model is equally applicable using an individual as the unit of analysis.

Variables in the "F" vector include the percentage of the cohort that is married and average number of dependents. The "I" vector includes student attributes such as average age and grade-point average. "TLFA" is comprised of variables such as the cohort's average tuition and the percentage of the cohort receiving financial aid. The "O" vector contains information on the undergraduate and graduate schools—their average quality, government spending on R&D, and the percentage of foreign baccalaureates enrolled—attended by persons in the cohort. "E" vector variables relate to the average starting salaries of new doctorates, the relation of doctorate to nondoctorate salaries, and the relation of salaries of new doctorates to those already in the field. Variables that capture the employment and unemployment experience of new and recent doctorates also are included. Since the cohort is the unit of analysis, variables on faculty-student interaction and social and academic integration are not included in the time-series version of this model.[8]

The Critical Role of Opportunity Costs

The model assumes opportunity costs affect student decisions that impact on TTD, but it does not explicitly allow for institutional decisions. From an economic perspective, a student's decision to undertake and complete a doctoral program involves a set of near-term costs in the form of opportunities foregone while the student pursues the doctorate. Current costs are borne in anticipation of future benefits, and both the costs and the future returns from the doctorate include monetary and nonmonetary elements.

There are at least three cost elements for graduate students, but they do not all affect TTD in the same way. As foregone earnings increase, TTD should decrease as pressure on students to enter the job market and earn an income rises. As foregone activities (e.g., work activities such as employment as a teaching assistant) increase, study time should decrease, with TTD increasing as a result. As financial outlays (primarily tuition) increase, incentives are created to finish

---

[8] Changes in programmatic requirements can elongate TTD: increasing the number of courses required for completion, requiring students to acquire additional competencies, lengthening time spent on doctorate-related research, and/or increasing the work experience that students must have to be eligible for the degree. Careful examination of these requirements would involve a separate study.

school as quickly as possible, and TTD decreases. Thus, student aid is expected to have both positive and negative effects on TTD, and those effects tend to be offsetting. As a result, the nature of the net effect (i.e., positive or negative) cannot be stated a priori.

Each vector in the model can be examined within the opportunity-cost framework. Other things equal, there will be a positive relationship between these costs and TTD. For example, it can be argued that married students have fewer costs associated with study time than single students. Since the opportunity costs of study time are often greater for single than married students, other things being equal, single students will spend less time on study and will have a high TTD. An opportunity-cost argument can also be made for the effects of family size. Other things remaining equal, as the number of dependents increases, the amount of time the student spends with the family also increases, causing TTD to rise. To the extent that women take primary responsibility for child rearing, married women with children will have a higher TTD than will married men with children.

Students who are better prepared to deal with the subject matter of their dissertation may find it less costly (in terms of time and effort required) to work on the doctorate. And it follows that students with an undergraduate degree in the same field as their doctoral study will, on average, have lower costs than those with a degree in a different field. Likewise, for those who enter graduate school with a high GRE score in the doctoral field, less time, and therefore less expense, probably will be needed to acquire the degree.

The effect of the quality of undergraduate education on TTD is not easy to assess. Study at a high-quality undergraduate institution may increase a student's preparation for graduate school, reducing the cost of pursuing the doctorate and resulting in faster progress to the degree. But attending a high-calibre institution can also lead to Davis' (1966) "frog pond" effect in which student expectations and grades drop, which in turn may increase TTD. The graduates of frog ponds may take their reduced expectations to graduate school, causing them to take longer to complete the doctorate.

Also not obvious is how quality of institution at the graduate level affects TTD. On the one hand, higher-quality institutions may provide their students with greater academic, social, and intellectual integration than lower-quality institutions and may be more efficient educators. Both phenomena reduce the costs of pursuing the doctorate, lower the costs of study time, and lower TTD. On the other hand, high-quality graduate schools may also impose more rigorous academic requirements on their doctoral candidates, requiring more research and study, with the ultimate effect of increasing TTD.

Market forces operate within the opportunity-cost context by determining what purchases the student foregoes while studying for a degree and what he or she can hope to receive in the future. The relationship between market forces and completion times is discussed later in this chapter.

Financial Aid and Its Impact on Completion Times

The Impact of Type of Aid

Because it offsets some of the income lost by being in school, financial aid in the form of fellowships, grants, and/or stipends reduces the opportunity costs of pursuing the doctorate. The pressure on students with aid to find outside employment is lessened, and they are freed to work on degree-related activities.

Aid that replaces a large amount of foregone income creates an incentive for some students to substitute leisure activities rather than study time for non-degree related work. Other students may increase TTD by enlarging the scope of their dissertation or taking an extra course or two. Although for many students financial aid is likely to increase their full-time work toward the doctorate, the net effect of fellowship aid on time to completion depends on whether students are more likely to partake in leisure or study activities when an award is made.[9]

Student behavior may also be affected if fellowship aid is contingent on a showing of successful progress toward the degree. The more stringent the criteria for demonstrating progress, the less likely students are to substitute leisure activities for study. However, it may be difficult to define "successful progress," since such criteria are fairly subjective (Prior, 1962).

Those with fellowships take less time to complete the degree than do recipients of teaching assistantships or those without aid, perhaps in part because they are more intellectually able. Students with teaching assistantships as their primary source of support have a lower opportunity cost for study than those who must support themselves through graduate school. Teaching assistants (TAs), because their aid package is dependent on the performance of services that take time away from doctorate-related activities, do not have as much time available for study as fellows or research assistants (RAs), suggesting TTD for the average TA will be longer than for the average fellow.

The situation with research assistantships is less clear. The wide range of duties assigned to RAs makes it difficult to generalize about the effects of such awards upon TTD. Those engaged in research related to their doctorate do not really give up study time when they spend job time in a way that facilitates

---

[9] For that matter, it also depends on whether the faculty who supervise dissertations have a preconceived idea of how long a dissertation should take, on university policies, and on curriculum matters.

completion of the doctorate. In contrast, those engaged in work unrelated to the doctorate may find that their job slows their progress toward the degree.

The opportunity-cost approach explains why students using their own earnings as the primary source of support are likely to take longer to complete the degree. A student employed full-time in a nondoctorate-related job must decide how to allocate non-work hours among leisure, study, family-related, and other activities. On average, the theory suggests, the time a working person devotes to doctoral study will be less than the time spent by those with teaching or research assistantships or fellowships.

A 1987 study by Abedi and Benkin found that mean TTD and mean RTD are over two years longer for students using their own earnings as a primary income source than for those whose money comes from other sources. But this same study found TTD was lower for students with "on-campus" earnings (including TAs and RAs) than for those with fellowships and grants. This finding is not consistent with the theory that TTD decreases as study costs decrease. To explore this discrepancy further, a separate mean TTD was computed for students reporting different primary sources of support (Table 3.1). In 4 of the 11 fields in 1986, and in 9 fields in 1987, fellowship recipients took less time to complete the degree than RAs. Likewise, in seven fields in both 1986 and 1987, fellows took less time to complete the doctorate than TAs. In nine fields in 1986 and 1987, students with research assistantships as their major source of support took less time to complete their doctorate than those with teaching assistantships. In all fields, students who used their own earnings as their primary source of income had substantially longer TTDs. For the 11 fields combined, fellows took less time to complete the doctorate than did TAs and those who used their own earnings to pay for school.

While the type of primary support mechanism does appear to affect TTD, this variable is not available for the 1967-1986 period and cannot be tested by modeling. However, models developed in Chapters 5 and 6 that assess the impact of any support from a given source reveal that financial aid has no consistent effect on TTD.

Effects on the Components of TTD

As noted earlier, the three components of TTD are time spent prior to graduate entrance (TPGE), registered time to the doctorate (RTD), and time not enrolled in the university (TNEU). In general, financial aid will reduce both TPGE and TNEU, but the effect of an increase in financial aid on RTD cannot be predicted with assurance, since it depends on the amount of foregone income replaced, the conditions under which aid is granted, and the form of aid received. Still, the expectation is that fellowships and dissertation-related research assistantships are more likely to lower RTD than nondoctorate-related research

TABLE 3.1: Mean Time to the Doctorate, by Primary Source of Support, 1986 and 1987

Field of Doctorate	Primary Source of Support: Teaching Assistantship 1986	Teaching Assistantship 1987	Research Assistantship 1986	Research Assistantship 1987	Fellowship Aid 1986	Fellowship Aid 1987	Own Earnings 1986	Own Earnings 1987
Total 11 Fields	8.77	8.68	7.94	8.16	8.49	8.07	13.56	13.91
Chemistry	7.50	6.69	6.37	6.47	6.87	5.75	10.83	11.93
Physics/Astronomy	8.64	8.49	7.30	7.61	8.09	6.90	13.70	11.85
Earth/Atmos/Marine Sciences	9.22	9.41	8.95	9.40	8.81	6.88	13.72	14.36
Math/Computer Sciences	7.74	8.90	7.92	8.67	8.57	8.92	13.83	13.75
Engineering	8.78	8.42	7.79	7.81	7.14	7.18	13.37	13.39
Agricultural Sciences	8.53	7.81	8.34	8.83	9.36	8.00	13.01	13.28
Biosciences	8.72	8.90	8.38	8.58	8.41	7.97	12.68	12.52
Health Sciences	10.93	9.78	8.96	10.96	10.95	9.83	14.75	15.54
Psychology	8.85	8.86	8.82	8.79	9.03	9.33	13.22	13.59
Economics	8.90	8.69	7.80	8.17	7.20	7.30	12.08	12.41
Social Sciences	10.66	10.80	11.29	10.06	10.36	9.85	15.27	15.70

SOURCE: National Research Council, Office of Scientific and Engineering Personnel, Doctorate Records File.

and teaching assistantships. Shortened TPGE occurs because financial aid makes it less costly for new graduates to pursue a graduate degree and increases the attractiveness of entering graduate school soon after completing an undergraduate degree.[10] Fellowships, which offer income without requiring a work commitment, are more likely to reduce TPGE than other forms of aid, particularly teaching assistantships and nondegree-related research assistantships. Whether one form of aid than another is more likely to affect TNEU will depend on its desirability relative to outside employment. For example, some students may prefer outside employment to teaching undergraduates.

Market Forces and Completion Times

The financial and other returns that students expect from completing a doctoral program can affect both their willingness to stay in school and TTD. The monetary incentive for earning a doctorate depends both on the absolute amount of the earnings expected and on the probability of employment. The returns from a given earnings stream and set of unemployment rates may be valued differently by students, depending on the importance they place on immediate versus future income and on their attitudes toward risk. Berger (1988) suggests that a single present-value measure can be used to incorporate expected returns into a model of student choice, but the analysis below assumes students consider expected earnings and the probability of unemployment separately.

Effects of Changes in Relative Salaries

Viewed from an opportunity-cost perspective, when starting salaries of new doctorates rise, income foregone by students while in graduate school increases. Increasing salaries increase the incentive for students to devote more time to completing doctorate and dissertation-related work, thus shortening TTD. The effect of salary on TTD may be partly offset, however, for Ph.D. candidates who get jobs before they finish their dissertation and are therefore likely to take longer to finish the doctorate.

An increase in the salary ratio of already employed doctorates to new or recent doctorates can mean different things. If postdoctoral experience is rewarded

---

10 The effect is two-fold. The student foregoes less income to attend graduate school and also has immediate access to a source of financial support. The latter is important for those who do not wish to borrow to finance their education and to those with a strong preference for current income.

such that salaries of experienced engineers are rising more rapidly than those of new entrants, for example, students have an incentive to complete their studies quickly. But if the increased salary ratio is the result of a poor market for new graduates, the signal is negative.

Changes in relative salaries also affect the three components of TTD:

RTD: When salaries for new doctorates rise, graduate students will generally find it worthwhile to shorten RTD by spending more time in study and dissertation-related activities. However, departmentally defined constraints may limit the amount by which students can reduce RTD.

TPGE: When the doctorate salary increases, TPGE is expected to shorten because the opportunity cost to the student of waiting to obtain the doctorate diminishes.

TNEU: TNEU is likely to fall when the starting salary of new doctorates rises relative to that of nondoctorates and when the salary of a doctorate with work experience rises relative to that of a new doctorate. A real rise in the starting salaries of doctorates will cause a decline in TNEU if the salary of a nondoctorate remains unchanged.

Effects of Employment Opportunities

Employment opportunities for new and recent doctorates are sometimes more visible and have greater impact on students than do relative salaries. Moreover, university placement offices are more likely to track the percentage of a graduating class with jobs than to compute the mean salaries of doctorates entering particular fields. The unemployment rate of new doctorates is an indicator of labor-market conditions and can be used in calculating future return for completing a doctoral program.

When employment opportunities increase for new and recent Ph.D.s, the opportunity costs increase to those remaining in graduate school. This creates an incentive for those working toward the doctorate to substitute degree-related work for leisure activity or outside employment, resulting in lower TTD. Conversely, when the opportunity cost of remaining in school falls, TTD for some students rises. The unemployment rates for new doctorates and for those without doctorates affect TTD in opposite directions. A rising unemployment rate for nondoctorates relative to the rate for doctorates increases the cost of remaining in graduate school, at least for those who either hold or plan to hold a non-university job, and will motivate students to finish the doctorate more rapidly. The percentage of students seeking employment or postdoctoral study will be used in lieu of unavailable unemployment data for new doctorates.

Effects on the Components of TTD

RTD: A rise in the unemployment rate for new doctorates leads to an increase in RTD and vice versa. An increase in unemployment among nondoctorates tends to lower RTD.

TPGE: A rise in unemployment for new doctorates increases TPGE, since it reduces apparent returns for earning a doctorate. But if the unemployment rate for nondoctorates rises relative to that for doctorates, TPGE will fall as the opportunity cost of attending graduate school is reduced.

TNEU: A rise in the unemployment rate for new doctorates encourages students to find and retain jobs prior to receipt of the doctorate, even if doing so lengthens TTD and TNEU. A rise in the nondoctorate unemployment rate, relative to the doctorate rate, reduces TNEU because it increases the benefits of obtaining the doctorate.

The Variables Used To Develop the Model

The primary source of the variables used in this study was the Doctorate Records File (DRF) maintained by the Office of Scientific and Engineering Personnel (OSEP) of the National Research Council. The DRF is a data base of doctorate recipients from U.S. universities spanning the period 1920 to the present. DRF data on TTD, RTD, TPGE, and TNEU for recent cohorts have been collected through the Survey of Earned Doctorates since 1958, although data on some of the variables became available more recently. OSEP also conducts the Survey of Doctorate Recipients (SDR), which provides biennial information on the employment status of scientific, engineering, and humanities doctorate holders.[11] Information in the SDR data base is used to construct market-force variables. Except where otherwise noted, the variables are for U.S. citizens and permanent residents. Altogether, 41 separate variables, falling into the 5 vectors of the study—family background, student attributes, tuition and financial aid, institutional environment, and market forces—are used.

---

11 A more complete description of this data base may be found in Betty D. Maxfield and Mary Belisle, *Science, Engineering, and Humanities Doctorates in the United States: 1983 Profile* (Washington, D.C.: National Academy Press, 1985).

In addition to data from the DRF and SDR, information on federal funding of students and universities, student scores, earnings in alternative employment, and unemployment were obtained from a variety of sources, including Battelle Columbus Laboratories (BCL), the Bureau of Labor Statistics (BLS), College Placement Council (CPC), the Educational Testing Service (ETS), Northwestern University's Endicott Report (ER), the National Science Foundation (NSF), and a number of professional associations such as the American Institute of Physics (AIP) and the Higher Education Research Institute (HERI). Data on classification of schools by research type came from the 1987 Carnegie Classification of Institutions of Higher Education. Data for "top 20" rankings came from the National Research Council's 1981-82 Assessment of Research Doctorate Programs in the United States, using NSF data in the Computer-Aided Science Policy and Research System (CASPAR).

4
THE RELATIONSHIP BETWEEN THE FIVE VECTORS OF VARIABLES AND TTD AND ITS COMPONENTS: A COMPARISON OF ZERO-ORDER CORRELATIONS

This chapter examines the zero-order correlations between a number of the independent variables used in the model and TTD and its component parts. The analysis discusses why the data are broken down by field and describes zero-order correlations between select variables in the five vectors in the model—family background, student attributes, tuition and financial aid, institutional environment, and market (economic and social) forces—and TTD and its components. It also provides correlations among the several salary and employment variables themselves and analyzes the amount of variation in TTD explained by each vector.

The Importance of Disaggregation by Field

Existing studies either addressed issues related to TTD aggregated over all fields (Wilson, 1965) or controlled for field differences using a set of dummy variables (Abedi and Benkin, 1987). The former approach ignores the possibility that a given independent variable (e.g., whether the student has an undergraduate degree in the same field) may have a different effect in some fields than in others, while the latter makes the rather stringent assumption that a one-unit change in an independent variable has the same effect on TTD for a student in chemistry, for example, as it does for a student in the biosciences. A number of studies of student aspirations and persistence suggest both assumptions are wrong (Biglan, 1973; Girves and Wemmerus, 1988; Thistlethwaite, 1962). And economic research suggests market conditions differ among scientific and engineering fields (Berger, 1988; Freeman, 1971). Failure to recognize that differences among field exist can give rise to models that give inaccurate explanations of why TTD changes.

The following sections provide field-specific data on the variables that the opportunity-cost analysis and the literature suggest have an effect on TTD. The zero-order correlations are suggestive, since the actual relationship between an independent variable and the dependent variable is captured by a model that tests their effect, *holding all other things equal*. These correlations highlight the differences among fields and help to explicate the interrelationship of the variables in each vector with TTD and its components. They also make it possible to examine the relationship between TTD and variables for which data are not available for sufficiently long periods of time.

Unless otherwise noted, all data are for the 1967-1986 period. All correlations are between the selected variable and TTD and its components. A single asterisk (*) denotes the correlation is significant at 1-percent level or greater. A double asterisk (**) denotes a significance level of 5 percent.

Changes in Family Background Characteristics

Of particular interest are the percentage of graduates in each doctorate cohort who are married (Table 4.1), the average number of dependents of doctorates in each cohort (Table 4.2), the percentage who are black (Table 4.3), the percentage who are Hispanic (Table 4.4), and the percentage who are women (Table 4.5).

Changes in Student Attributes

The variables of interest are average age of the cohort at the start of the doctoral program (Table 4.6), percentage in the cohort who attended a highly selective undergraduate school (Table 4.7), and percentage of the cohort with an undergraduate degree in the same field as their doctorate (Table 4.8).

For the reasons discussed in Chapter 3, a thorough analysis of TTD should employ a measure either of student ability, such as undergraduate or graduate cumulative grade-point average,[12] or of achievement level, using scores from the SAT, ACT, or GRE. Unfortunately, the DRF does not contain data either on student grades or on predoctorate test scores. To develop a "proxy" measure of the skills that a given cohort possesses, we used a variable equal to the percentage of new doctorates in each cohort who attended a selective undergraduate institution, where the average incoming 1973-74 freshmen earned a combined SAT verbal and math score of 1,300 or higher. The assumption is that the larger the percentage of students from institutions of this type, the larger the overall ability level of the students in a given cohort.

---

[12] Student grades pose a technical problem when they are aggregated at the cohort level because the ordinal scales used to grade students at different institutions are not additive.

TABLE 4.1: Correlation for Percent Married, 1967-1986

Field	TTD	RTD	TPGE	TNEU
Chemistry	-0.46**	-0.67*	-0.37	0.80*
Physics/Astronomy	-0.68*	-0.82*	-0.59*	0.57*
Earth/Atmospheric/Marine Sciences	-0.44**	-0.89*	-0.08	0.84*
Mathematics/Computer Sciences	-0.80*	-0.87*	-0.74*	-0.29
Engineering	-0.69*	-0.80*	-0.11	0.05
Agricultural Sciences	-0.07*	-0.86*	-0.63*	0.68*
Biosciences	-0.62*	-0.92*	-0.60*	0.78*
Health Sciences	-0.66*	-0.80*	-0.48**	-0.41
Psychology	-0.82*	-0.85*	-0.85*	-0.38
Economics	-0.66*	-0.91*	-0.04	0.68*
Social Sciences	-0.77*	-0.93*	-0.45**	0.65*

NOTE: These are zero-order correlation coefficients.
*Denotes correlation is statistically significant at 1% level or greater.
**Denotes correlation is significant at 5% level or greater.

TABLE 4.2: Correlation for Average Number of Dependents, 1967-1986

Field	TTD	RTD	TPGE	TNEU
Chemistry	-0.66*	-0.84*	-0.42	0.76*
Physics/Astronomy	-0.83*	-0.91*	-0.64*	0.46**
Earth/Atmospheric/Marine Sciences	-0.48**	-0.90*	-0.03	0.85*
Mathematics/Computer Sciences	-0.84*	-0.91*	-0.74*	-0.36
Engineering	-0.85*	-0.90*	-0.15	-0.11
Agricultural Sciences	0.04	-0.88*	0.72*	0.78*
Biosciences	-0.50**	-0.88*	-0.46**	0.85*
Health Sciences	-0.58*	-0.74*	-0.48**	-0.29
Psychology	-0.74*	-0.81*	-0.78*	-0.24
Economics	-0.47**	-0.96*	0.18	0.85*
Social Sciences	-0.71*	-0.92*	-0.37	0.75*

NOTE: These are zero-order correlation coefficients.
*Denotes correlation is statistically significant at 1% level or greater.
**Denotes correlation is significant at 5% level or greater.

TABLE 4.3: Correlation for Percent Black, 1974-1986

Field	TTD	RTD	TPGE	TNEU
Chemistry	0.09	0.008	-0.20	0.33
Physics/Astronomy	0.05	0.49	0.24	-0.48
Earth/Atmospheric/Marine Sciences	-0.007	0.25	-0.05	-0.59
Mathematics/Computer Sciences	-0.20	-0.19	-0.27	-0.03
Engineering	0.25	0.47	-0.37	0.03
Agricultural Sciences	0.16	0.33	0.01	-0.13
Biosciences	-0.09	-0.15	-0.15	0.33
Health Sciences	0.01	0.06	-0.15	0.08
Psychology	0.57*	0.62**	0.53	0.46
Economics	0.60*	0.74*	0.55**	-0.93*
Social Sciences	0.70*	0.75*	0.61**	0.10

NOTE: These are zero-order correlation coefficients.
*Denotes correlation is statistically significant at 1% level or greater.
**Denotes correlation is significant at 5% level or greater.

TABLE 4.4: Correlation for Percent Hispanic, 1974-1986

Field	TTD	RTD	TPGE	TNEU
Chemistry	0.16	0.54	-0.14	-0.33
Physics/Astronomy	0.07	0.25	0.53**	-0.40
Earth/Atmospheric/Marine Sciences	0.22	0.34	-0.03	-0.32
Mathematics/Computer Sciences	0.79*	0.73*	0.73*	0.53
Engineering	0.05	0.30	-0.49	-0.41
Agricultural Sciences	0.22	0.60**	-0.26	-0.22
Biosciences	0.84*	0.86*	0.87*	-0.17
Health Sciences	0.66*	0.62**	0.71*	0.52
Psychology	0.93*	0.93*	0.92*	0.91*
Economics	0.68*	0.75*	0.41	-0.53
Social Sciences	0.89*	0.93*	0.83*	0.31

NOTE: These are zero-order correlation coefficients.
*Denotes correlation is statistically significant at 1% level or greater.
**Denotes correlation is significant at 5% level or greater.

TABLE 4.5: Correlation for Percent Female, 1967-1986

Field	TTD	RTD	TPGE	TNEU
Chemistry	0.63*	0.78*	0.48**	-0.70*
Physics/Astronomy	0.68*	0.82*	0.69*	-0.58*
Earth/Atmospheric/Marine Sciences	0.59*	0.92*	0.06	-0.76*
Mathematics/Computer Sciences	0.93*	0.97*	0.86*	0.46**
Engineering	0.65*	0.72*	-0.07	0.04
Agricultural Sciences	0.08	0.87*	-0.67*	-0.67*
Biosciences	0.70*	0.96*	0.68*	-0.72*
Health Sciences	0.82*	0.88*	0.71*	0.58*
Psychology	0.91*	0.94*	0.93*	0.48**
Economics	0.63*	0.91*	0.03	-0.70*
Social Sciences	0.86*	0.96*	0.56*	-0.57*

NOTE: These are zero-order correlation coefficients.
*Denotes correlation is statistically significant at 1% level or greater.
**Denotes correlation is significant at 5% level or greater.

TABLE 4.6: Correlation for Average Age at Start of Doctoral Program, 1967-1986.

Field	TTD	RTD	TPGE	TNEU
Chemistry	0.85*	0.89*	0.73*	-0.51*
Physics/Astronomy	0.62*	0.61*	0.78*	0.23
Earth/Atmospheric/Marine Sciences	0.31	-0.16	0.49**	0.49**
Mathematics/Computer Sciences	0.93*	0.88*	0.91*	0.65*
Engineering	0.29	0.08	0.83*	0.42
Agricultural Sciences	0.44**	-0.45**	0.75*	0.70*
Biosciences	0.90*	0.77*	0.91*	-0.11
Health Sciences	0.76*	0.61*	0.88*	0.65*
Psychology	0.99*	0.96*	0.98*	0.76*
Economics	0.72*	-0.04	0.76*	0.42
Social Sciences	0.89*	0.61*	0.93*	0.19

NOTE: These are zero-order correlation coefficients.
*Denotes correlation is statistically significant at 1% level or greater.
**Denotes correlation is significant at 5% level or greater.

Table 4.7: Correlation for Percent from Selective Undergraduate Schools, 1967-1986

Field	TTD	RTD	TPGE	TNEU
Chemistry	-0.61*	-0.62*	-0.40	0.27
Physics/Astronomy	-0.55*	-0.43	-0.44**	-0.31
Earth/Atmospheric/Marine Sciences	-0.30	-0.13	-0.29	-0.11
Mathematics/Computer Sciences	0.01	-0.003	0.11	-0.13
Engineering	-0.25	-0.22	-0.37	-0.02
Agricultural Sciences	-0.32	0.60*	-0.56*	-0.80*
Biosciences	0.30	0.19	0.37	0.01
Health Sciences	-0.06	-0.13	-0.30	0.15
Psychology	-0.39	-0.47**	-0.42	0.02
Economics	0.11	0.11	0.08	-0.11
Social Sciences	-0.70*	-0.90*	-0.34	0.68*

NOTE: These are zero-order correlation coefficients.
*Denotes correlation is statistically significant at 1% level or greater.
**Denotes correlation is significant at 5% level or greater.

TABLE 4.8: Correlation for Percent with Undergraduate Degree in Doctoral Field, 1974-1986

Field	TTD	RTD	TPGE	TNEU
Chemistry	-0.002	-0.06	0.07	0.22
Physics/Astronomy	0.76*	0.84*	0.45**	-0.26
Earth/Atmospheric/Marine Sciences	-0.29	0.67*	-0.05	0.68*
Mathematics/Computer Sciences	-0.82*	-0.80*	-0.83*	-0.40
Engineering	-0.80*	-0.89*	-0.07	-0.02
Agricultural Sciences	0.28	-0.70*	0.65*	0.85*
Biosciences	0.52*	0.89*	0.50*	-0.85*
Health Sciences	0.77	0.83*	0.66*	0.57*
Psychology	0.07	0.16	0.08	-0.16
Economics	0.24	0.85*	-0.27	-0.85*
Social Sciences	0.57*	0.79*	0.23	-0.69*

NOTE: These are zero-order correlation coefficients.
*Denotes correlation is statistically significant at 1% level or greater.
**Denotes correlation is significant at 5% level or greater.

Changes in Tuition and Financial Aid

The variables of interest here are average tuition and fees paid in a given year (Table 4.9); percentage of students with federal support (Table 4.10), private foundation support (Table 4.11), research assistantships (Table 4.12), or teaching assistantships (Table 4.13); and percentage of students who relied on their own earnings as their primary means of support (Table 4.14).

Changes in Institutional Environment

This subsection examines the relationship between a select number of aggregate measures of institutional environment and TTD. These are the percentage of students with a baccalaureate from a foreign institution (Table 4.15, p. 54), the ratio of full-time equivalent faculty to doctorate recipients (Table 4.16, p. 55), the ratio of the dollar value of government R&D

TABLE 4.9: Correlation for Average Tuition Paid

Field	TTD	RTD	TPGE	TNEU
Chemistry	0.07	-0.05	0.05	0.33
Physics/Astronomy	-0.007	-0.16	0.12	0.38
Earth/Atmospheric/Marine Sciences	-0.05	-0.30	-0.05	0.40
Mathematics/Computer Sciences	-0.35	-0.31	-0.36	-0.17
Engineering	-0.14	-0.15	0.27	-0.10
Agricultural Sciences	0.28	-0.17	0.17	0.36
Biosciences	-0.44**	-0.39	-0.46**	0.09
Health Sciences	-0.26	-0.53**	0.16	0.21
Psychology	-0.45**	-0.40	-0.46**	-0.52*
Economics	-0.54**	-0.27	-0.18	-0.06
Social Sciences	0.42	-0.33	-0.54*	0.02

NOTES: (1) These are zero-order correlation coefficients. (2) Tuition lagged three years. Weights were used to aggregate public and private institutions. Since national averages are not available for graduate tuition and fees, our analysis assumes that undergraduate tuition is a good proxy variable. The assumption is that undergraduate and graduate tuitions are highly correlated and that increases in the former are accompanied by similar increases in the latter.
*Denotes correlation is statistically significant at 1% level or greater.
**Denotes correlation is significant at 5% level or greater.

TABLE 4.10: Correlation for Percent with Primary Support from Federal Government, 1967-1986

Field	TTD	RTD	TPGE	TNEU
Chemistry	-0.52**	-0.70*	-0.44	0.76*
Physics/Astronomy	-0.51*	-0.67*	-0.51*	0.64*
Earth/Atmospheric/Marine Sciences	-0.49**	-0.83*	-0.04	0.74*
Mathematics/Computer Sciences	-0.87*	-0.91*	-0.81*	-0.38
Engineering	-0.61*	-0.69*	0.14	-0.04
Agricultural Sciences	-0.10	-0.59*	0.47**	0.39
Biosciences	-0.69*	-0.94*	-0.67*	0.72*
Health Sciences	-0.74*	-0.86*	-0.48**	-0.52**
Psychology	-0.94*	-0.94*	-0.95*	-0.60*
Economics	-0.91*	-0.60*	-0.47**	0.20
Social Sciences	-0.88*	-0.67*	-0.87*	-0.07

NOTE: These are zero-order correlation coefficients.
*Denotes correlation is statistically significant at 1% level or greater.
**Denotes correlation is significant at 5% level or greater.

TABLE 4.11: Correlation for Percent with Primary Support from Private Foundations, 1967-1986

Field	TTD	RTD	TPGE	TNEU
Chemistry	-0.14	-0.18	0.06	0.05
Physics/Astronomy	-0.58*	-0.56*	-0.21	-0.07
Earth/Atmospheric/Marine Sciences	0.01	-0.26	0.18	0.40
Mathematics/Computer Sciences	-0.06	-0.23	0.03	0.26
Engineering	-0.26	-0.38	-0.33	0.31
Agricultural Sciences	0.43	0.30	0.02	0.09
Biosciences	0.61*	0.18	0.59*	0.51**
Health Sciences	-0.16	-0.24	-0.40	0.10
Psychology	-0.56*	-0.65*	-0.62*	-0.07
Economics	-0.02	-0.78*	0.43	0.86*
Social Sciences	-0.21	-0.61*	0.24	0.90*

NOTE: These are zero-order correlation coefficients.
*Denotes correlation is statistically significant at 1% level or greater.
**Denotes correlation is significant at 5% level or greater.

TABLE 4.12: Correlation for Percent with Primary Support from Research Assistantships, 1967-1986

Field	TTD	RTD	TPGE	TNEU
Chemistry	0.63*	0.76*	0.46**	-0.65*
Physics/Astronomy	0.60*	0.73*	0.56*	-0.54*
Earth/Atmospheric/Marine Sciences	0.33	0.82*	-0.09	-0.88*
Mathematics/Computer Sciences	0.72*	0.75*	0.72*	0.23
Engineering	0.79*	0.80*	0.09	-0.06
Agricultural Sciences	-0.01	0.85*	-0.66*	-0.77*
Biosciences	0.77*	0.96*	0.74*	-0.63*
Health Sciences	0.80*	0.91*	0.62*	0.58*
Psychology	-0.64*	0.66*	-0.73*	-0.29
Economics	0.41	0.67*	-0.09	-0.59*
Social Sciences	0.71*	0.77*	0.43	-0.39

NOTE: These are zero-order correlation coefficients.
*Denotes correlation is statistically significant at 1% level or greater.
**Denotes correlation is significant at 5% level or greater.

TABLE 4.13: Correlation for Percent with Primary Support from Teaching Assistantships, 1967-1986

Field	TTD	RTD	TPGE	TNEU
Chemistry	0.68*	0.72*	0.46**	-0.38
Physics/Astronomy	0.87*	0.88*	0.48**	-0.03
Earth/Atmospheric/Marine Sciences	-0.54*	-0.61*	-0.24	0.40
Mathematics/Computer Sciences	0.45	0.56*	0.31	0.15
Engineering	0.66*	0.69*	0.43	0.09
Agricultural Sciences	-0.02	0.65*	-0.46**	-0.61*
Biosciences	0.17	0.64*	0.10	-0.85*
Health Sciences	0.20	0.14	0.39	0.07
Psychology	0.04	0.10	0.04	-0.18
Economics	0.24	0.86*	-0.29	-0.86*
Social Sciences	0.61*	0.81*	0.24	-0.65

NOTE: These are zero-order correlation coefficients.
*Denotes correlation is statistically significant at 1% level or greater.
**Denotes correlation is significant at 5% level or greater.

TABLE 4.14: Correlation for Percent with Primary Support from Own Earnings, 1977-1986

Field	TTD	RTD	TPGE	TNEU
Chemistry	0.63**	0.77*	0.11	0.25
Physics/Astronomy	0.52	-0.16	0.47	0.62
Earth/Atmospheric/Marine Sciences	0.71**	0.81*	0.40	0.14
Mathematics/Computer Sciences	0.81*	0.69**	0.72**	0.79*
Engineering	-0.32	-0.32	-0.33	-0.08
Agricultural Sciences	0.66**	0.31	0.49	0.77*
Biosciences	0.79*	0.73**	0.78*	0.63**
Health Sciences	0.92*	0.89*	0.93*	0.85*
Psychology	0.84*	0.85*	0.81*	0.78*
Economics	0.66**	0.79*	0.05	-0.46
Social Sciences	0.91*	0.92*	0.84*	0.86*

NOTE: These are zero-order correlation coefficients.
*Denotes correlation is statistically significant at 1% level or greater.
**Denotes correlation is significant at 5% level or greater.

TABLE 4.15: Correlation for Percent with Baccalaureate from Foreign Institutions, 1967-1986

Field	TTD	RTD	TPGE	TNEU
Chemistry	0.42	0.34	0.14	0.19
Physics/Astronomy	0.32	0.17	0.07	0.53*
Earth/Atmospheric/Marine Sciences	0.22	-0.13	0.40	0.41
Mathematics/Computer Sciences	0.90*	0.90*	0.88*	0.50*
Engineering	0.76*	0.74*	0.70*	0.16
Agricultural Sciences	0.21	-0.29	0.32	0.41
Biosciences	-0.52**	-0.69*	-0.53*	0.51*
Health Sciences	-0.46**	-0.45**	-0.18	-0.48**
Psychology	-0.28	-0.27	-0.34	-0.09
Economics	0.68*	0.53*	0.35	-0.29
Social Sciences	-0.32	-0.42	0.19	0.38

NOTE: These are zero-order correlation coefficients.
*Denotes correlation is statistically significant at 1% level or greater.
**Denotes correlation is significant at 5% level or greater.

TABLE 4.16: Correlation for Number of Full-Time Equivalent Faculty

Field	TTD	RTD	TPGE	TNEU
Chemistry	0.68*	0.85*	0.37	-0.65*
Physics/Astronomy	0.88*	0.95*	0.65*	-0.33
Earth/Atmospheric/Marine Sciences	0.38	0.82*	-0.09	-0.81*
Mathematics/Computer Sciences	0.80*	0.90*	0.69*	0.33
Engineering	0.79*	0.93*	0.19	-0.14
Agricultural Sciences	-0.06	0.83*	-0.66*	-0.79*
Biosciences	0.28	0.75*	0.25	-0.91*
Health Sciences	0.48	0.60*	0.49	0.19
Psychology	0.60*	0.69*	0.64*	0.04*
Economics	0.22	0.90*	-0.37	-0.92*
Social Sciences	0.52**	0.83*	0.11	-0.85*

NOTES: (1) These are zero-order correlation coefficients. (2) Period for TTD is 1967-1986; FACULTY, a crude proxy for the number of mentors available to doctorate students, is lagged, 1964-1983.
*Denotes correlation is statistically significant at 1 percent level or greater.
**Denotes correlation is significant at 5 percent level or greater.

TABLE 4.17: Correlation for Government R&D Spending

Field	TTD	RTD	TPGE	TNEU
Chemistry	0.54*	0.70*	0.47*	-0.72*
Physics/Astronomy	0.59*	0.73*	0.64*	-0.61*
Earth/Atmospheric/Marine Sciences	0.63*	0.92*	0.07	-0.72*
Mathematics/Computer Sciences	0.92*	0.91*	0.90*	0.49**
Engineering	0.62*	0.71*	-0.14	0.01
Agricultural Sciences	0.18	0.86*	-0.56*	-0.60*
Biosciences	0.80*	0.98*	0.79*	-0.63*
Health Sciences	0.89*	0.95*	0.67*	0.70*
Psychology	0.97*	0.98*	0.97*	0.63*
Economics	0.75*	0.89*	0.14	-0.61*
Social Sciences	0.95	0.96*	0.71*	-0.40

NOTES: (1) These are zero-order correlation coefficients. (2) Period for TTD is 1967-1986 and for R&D is 1964-1983.
*Denotes correlation is statistically significant at 1 percent level or greater.
**Denotes correlation is significant at 5 percent level or greater.

expenditures to doctorate recipients (Table 4.17), the percentage of doctorate recipients who received an undergraduate degree from a Research I school as identified by the Carnegie Classification (Table 4.18), the percentage of students who received an undergraduate degree from a "top 40" school as identified by the NRC's Assessment of Research-Doctorate Programs in the United States (Table 4.19), the percentage of students who received a graduate degree from a Carnegie-classified Research I or Research II school (Table 4.20), and the percentage of students who received a graduate degree from a "top 40" school (Table 4.21).

Changes in Market Forces

Salary Variables

An exhaustive review of salary data revealed differences in the quantity and quality of various sources (Tables 4.22-4.25). Only seven data files were used; others were excluded either because their academic field classifications were incompatible with those in this study or because the time spans of data collection were inadequate.[13]

---

[13] The sources for data on salary were the following: the American Institute of Physics, baccalaureate salary data beginning in 1965 for physics and astronomy [three missing years of data (1964, 1966, and 1967) were generated using an instrumental variable based on the Endicott Report data for physics]; Battelle Columbus Laboratories' data series that begins in 1968 for baccalaureate and doctorate salaries in engineering, chemistry, and physics (BCL's data series for life sciences was considered too aggregated for use in the model, but the data are shown in the correlation table with SDR salary for biological scientists); College Placement Council data on salary offers to baccalaureates, starting in 1964 for chemistry and math (excluding computer sciences); Endicott Report data on baccalaureate salary starting in 1964 for chemistry, math, engineering, and the combined field of economics and finance; the Survey of Doctorate Recipients, the only source of doctorate salary data for all 11 fields (such data have been collected on a biennial basis since 1973; however, an instrumental variable was created based on Bureau of Labor Statistics' Weekly Earnings data to provide even-year data and to project salaries back to 1964); and baccalaureate salary data from the National Survey of Hospital and Medical School Salaries starting in 1964 for staff nurses, used as a proxy for health sciences.

TABLE 4.18: Correlation for Percent with Baccalaureate from Category I Research University, 1967-1986

Field	TTD	RTD	TPGE	TNEU
Chemistry	-0.34	-0.37	0.26	-0.14
Physics/Astronomy	0.26	0.42	0.31	-0.58*
Earth/Atmospheric/Marine Sciences	-0.34	-0.22	-0.22	-0.03
Mathematics/Computer Sciences	0.31	0.30	-0.38	0.02
Engineering	-0.14	-0.03	-0.31	-0.19
Agricultural Sciences	0.17	-0.35	0.26	0.50**
Biosciences	0.58*	0.82*	0.58*	-0.68*
Health Sciences	0.04	-0.08	-0.10	0.14
Psychology	-0.49**	-0.50**	-0.48**	-0.35
Economics	0.45**	-0.27	0.51**	0.52**
Social Sciences	-0.40	-0.67*	-0.10	0.72*

NOTES: These are zero-order correlation coefficients. Category I Research University is taken from the Carnegie Classification of Colleges and Universities.
*Denotes correlation is statistically significant at 1% level or greater.
**Denotes correlation is significant at 5% level or greater.

TABLE 4.19: Correlation for Percent with Baccalaureate from "Top 40" School, 1967-1986

Field	TTD	RTD	TPGE	TNEU
Chemistry	-0.63*	-0.75*	-0.13	0.36
Physics/Astronomy	0.01	0.12	0.06	-0.40
Earth/Atmospheric/Marine Sciences	0.04	0.12	0.06	-0.22
Mathematics/Computer Sciences	-0.10	-0.08	-0.08	-0.16
Engineering	-0.79*	-0.78*	-0.59*	-0.15
Agricultural Sciences	-0.39	0.06	-0.25	-0.35
Biosciences	0.54*	0.81*	0.54*	-0.71*
Health Sciences	0.03	-0.05	-0.18	0.19
Psychology	-0.55*	-0.61*	-0.57*	-0.20
Economics	0.30	-0.11	0.21	0.24
Social Sciences	-0.41	-0.65*	-0.12	0.65*

NOTES: These are zero-order correlation coefficients. "Top 40" refers to those schools so identified in the NRC's Assessment of Research Doctorate Programs in the United States.
*Denotes correlation is statistically significant at 1% level or greater.
**Denotes correlation is significant at 5% level or greater.

TABLE 4.20: Correlation for Percent with Graduate Degree from Category I or Category II Research School, 1967-1986

Field	TTD	RTD	TPGE	TNEU
Chemistry	0.03	0.20	0.17	-0.64*
Physics/Astronomy	0.13	0.27	0.25	-0.59*
Earth/Atmospheric/Marine Sciences	-0.46**	-0.79*	-0.02	0.66*
Mathematics/Computer Sciences	-0.17	-0.13	-0.13	-0.38
Engineering	-0.14	-0.06	-0.53*	-0.08
Agricultural Sciences	-0.17	-0.47**	0.18	0.29
Biosciences	0.003	-0.44**	0.04	0.74*
Health Sciences	-0.15	-0.33	-0.23	0.06
Psychology	-0.79*	-0.85*	-0.82*	-0.26
Economics	0.17	-0.63*	0.67*	0.77*
Social Sciences	-0.68*	-0.87*	-0.38	0.67*

NOTES: These are zero-order correlation coefficients. Category I Research University is taken from the Carnegie Classification of Colleges and Universities.
*Denotes correlation is statistically significant at 1% level or greater.
**Denotes correlation is significant at 5% level or greater.

TABLE 4.21: Correlation for Percent with Graduate Degree from "Top 40" School, 1967-1986

Field	TTD	RTD	TPGE	TNEU
Chemistry	-0.16	0.01	0.02	-0.58*
Physics/Astronomy	0.17	0.32	0.25	-0.58*
Earth/Atmospheric/Marine Sciences	0.005	0.14	-0.29	-0.20
Mathematics/Computer Sciences	0.08	0.10	0.12	-0.19
Engineering	-0.46**	-0.47**	-0.50**	-0.02
Agricultural Sciences	-0.28	-0.82*	0.26	0.53*
Biosciences	0.05	-0.29	0.05	0.56*
Health Sciences	-0.08	-0.28	-0.17	0.16
Psychology	-0.80*	-0.87*	-0.84*	-0.29
Economics	0.35	-0.40	0.72*	0.59*
Social Sciences	-0.45*	-0.73*	-0.09	0.77*

NOTES: These are zero-order correlation coefficients. "Top 40" refers to those schools so identified in the NRC's Assessment of Research Doctorate Programs in the United States.
*Denotes correlation is statistically significant at 1% level or greater.
**Denotes correlation is significant at 5% level or greater.

TABLE 4.22: Correlation for Average Salary of Recent Doctorate Recipients

Field	TTD	RTD	TPGE	TNEU
Chemistry	0.70*	0.79*	0.40	-0.50**
Physics/Astronomy	0.67*	0.62*	0.25	0.23
Earth/Atmospheric/Marine Sciences	-0.57*	-0.52**	-0.28	0.19
Mathematics/Computer Sciences	0.49**	0.59*	0.35	0.24
Engineering	0.78*	0.84*	0.40	0.02
Agricultural Sciences	-0.20	-0.86*	0.57*	0.57*
Biosciences	-0.73*	-0.96*	-0.73*	0.70*
Health Sciences	-0.80*	-0.82*	-0.60*	-0.70*
Psychology	-0.96*	-0.97*	-0.97*	-0.60*
Economics	-0.68*	-0.37	-0.53*	0.05
Social Sciences	-0.89*	-0.97*	-0.63*	0.52**

NOTES: (1) These are zero-order correlation coefficients. (2) Specifically, SDR salary is regressed on weekly earnings, and the coefficients from this regression are used to estimate salaries in the missing years; SDRSAL is lagged three years.
*Denotes correlation is statistically significant at 1% level or greater.
**Denotes correlation is significant at 5% level or greater.

TABLE 4.23: Correlation Between SALRAT1 and TTD and Its Components

Field	TTD	RTD	TPGE	TNEU
Chemistry	-0.69*	-0.87*	-0.43	0.71*
Physics/Astronomy	-0.59*	-0.75*	-0.52*	0.54*
Earth/Atmospheric/Marine Sciences	-0.007	0.10	-0.002	-0.13
Mathematics/Computer Sciences	-0.87*	-0.95*	-0.78*	-0.37
Engineering	-0.75*	-0.85*	0.08	0.04
Agricultural Sciences	0.18	0.66*	-0.37*	-0.47**
Biosciences	0.60*	0.91*	0.57*	-0.77*
Health Sciences	0.38	0.34	0.42	0.32
Psychology	0.76*	0.76*	0.80*	0.44**
Economics	0.41	0.56*	0.20	-0.39
Social Sciences	0.23	0.29	0.19	-0.25

NOTES: (1) These are zero-order correlation coefficients. (2) The years prior to 1973 are projected; SALRAT1 is lagged three years.
*Denotes correlation is statistically significant at 1% level or greater.
**Denotes correlation is significant at 5% level or greater.

TABLE 4.24: Correlation for Salary Ratio of Doctorates 10 Years After Degree

Field	TTD	RTD	TPGE	TNEU
Chemistry	0.63	0.72*	0.51**	-0.56*
Physics/Astronomy	0.62	0.57*	0.34	0.14
Earth/Atmospheric/Marine Sciences	-	-	-	-
Mathematics/Computer Sciences	0.56	0.63*	0.42	0.33
Engineering	0.55	0.63*	0.13	0.01
Agricultural Sciences	-	-	-	-
Biosciences	-	-	-	-
Health Sciences	-0.51	-0.61*	-0.56*	-0.26
Psychology	-	-	-	-
Economics	0.21	0.07	-0.01	-0.01
Social Sciences	-	-	-	-

NOTES: (1) These are zero-order correlation coefficients. (2) A comparison is made to the baccalaureate rather than the master's salary because of the larger number of observations in the former category; SALRAT10 is lagged three years.
*Denotes correlation is statistically significant at 1% level or greater.
**Denotes correlation is significant at 5% level or greater.

Employment Indicator Variables

The percentage of doctorate recipients seeking postgraduate employment is a reliable indicator of job market opportunity. Job opportunities, in turn, are likely to affect TTD. Data on job-seeking behavior are easy to obtain and reasonably reliable (Tables 4.26-4.30). However, because such data are collected at the time the doctoral candidate is completing the degree, they may understate employment prospects, because finding a job after graduation takes time. Data on job-seeking activity have been used in studies by Freeman (1971).

The Stock Variable

The zero-order correlations between TTD and its components and the number of doctorates in the United States divided by the U.S. population are shown in Table 4.31.

TABLE 4.25: Correlation Between SDR Salaries and Salaries Reported by Other Sources, 1968-1986

Doctorate Field	SDRSAL10 W/BCPCREAL	SDRSAL10 W/BSALREAL	SDRSAL10 W/BSALPROF	SDRSAL10 W/BATTELLE1	SDRSAL10 W/BATTELLE2
Chemistry	0.46**	0.36	-	0.89*	0.87*
Physics/Astronomy	-	-	0.14	0.57	0.60*
Earth/Atmospheric/Marine Sci.	-	-	-	-	-
Math/Computer Sciences	0.24	-0.08	-	-	-
Engineering	-	-0.23	-	0.55*	0.61*
Agricultural Sciciences	-	-	-	-	-
Biosciences	-	-	-	0.88*	0.88*
Health Sciences	-	-	0.69*	-	-
Psychology	-	-	-	-	-
Economics	-	0.28	-	-	-
Social Sciences	-	-	-	-	-

NOTES: (1) These are zero-order correlation coefficients. (2) Battelle data on life sciences were correlated with data on biosciences because a separate biosciences series did not exist. (3) Acronyms are defined in Appendix B, pp. 175-177.

*Denotes correlation is statistically significant at 1% level or better.

**Denotes correlation is significant at 5% level or better.

TABLE 4.26: Correlation for Percent Seeking Postgraduate Employment

Field	TTD	RTD	TPGE	TNEU
Chemistry	0.62*	0.58*	0.44**	-0.17
Physics/Astronomy	0.75*	0.60*	0.36	0.44**
Earth/Atmospheric/Marine Sciences	-0.02	0.42	-0.12	-0.60*
Mathematics/Computer Sciences	0.30	0.41	0.12	0.11
Engineering	0.65*	0.62*	0.55*	0.24
Agricultural Sciences	0.20	0.80*	-0.39*	-0.60*
Biosciences	0.24	0.61*	0.19	-0.71*
Health Sciences	0.51*	0.55*	0.59*	0.27
Psychology	0.74*	0.78*	0.78*	0.30
Economics	0.19	0.75*	-0.33	-0.72*
Social Sciences	0.84*	0.95*	0.54*	-0.58*

NOTES: (1) These are zero-order correlation coefficients. (2) SEEK variable is lagged three years.
*Denotes correlation is statistically significant at 1% level or greater.
**Denotes correlation is significant at 5% level or greater.

TABLE 4.27: Correlation for Percent with Definite Employment or Postdoctoral Appointment

Field	TTD	RTD	TPGE	TNEU
Chemistry	-0.64*	-0.61*	-0.45**	0.19
Physics/Astronomy	-0.75*	-0.61*	-0.38	-0.37
Earth/Atmospheric/Marine Sciences	-0.02	-0.44**	0.15	0.60*
Mathematics/Computer Sciences	-0.31	-0.43	-0.14	-0.09
Engineering	-0.63*	-0.60*	-0.53*	-0.24
Agricultural Sciences	-0.20	-0.77*	0.37	0.56*
Biosciences	-0.18	-0.54*	-0.11	0.68*
Health Sciences	-0.62*	-0.67*	-0.62*	-0.36
Psychology	-0.71*	-0.75*	-0.75*	-0.26
Economics	-0.09	-0.66*	0.33	0.66*
Social Sciences	-0.82*	-0.95*	-0.52*	0.60*

NOTES: (1) These are zero-order correlation coefficients. (2) DEFIN variable is lagged three years.
*Denotes correlation is statistically significant at 1% level or greater.
**Denotes correlation is significant at 5% level or greater.

TABLE 4.28: Correlation for Overall U.S. Unemployment Rate

Field	TTD	RTD	TPGE	TNEU
Chemistry	0.55*	0.61*	0.61*	-0.55*
Physics/Astronomy	0.61*	0.64*	0.63*	-0.28
Earth/Atmospheric/Marine Sciences	0.39	0.73*	0.15	-0.72*
Mathematics/Computer Sciences	0.81*	0.77*	0.73*	0.57*
Engineering	0.54*	0.63*	-0.15	0.06
Agricultural Sciences	0.18	0.71*	-0.37	-0.51*
Biosciences	0.71*	0.82*	0.69*	-0.47*
Health Sciences	0.72*	0.81*	0.50**	0.53*
Psychology	0.78*	0.78*	0.82*	0.52*
Economics	0.61*	0.71*	0.07	-0.44**
Social Sciences	0.77*	0.78*	0.63*	-0.34*

NOTES: (1) These are zero-order correlation coefficients. (2) Unemployment variable is lagged three years.
*Denotes correlation is statistically significant at 1% level or greater.
**Denotes correlation is significant at 5% level or greater.

TABLE 4.29: Correlation for Unemployment Rate of College-Educated Population

Field	TTD	RTD	TPGE	TNEU
Chemistry	0.58*	0.63*	0.63*	-0.57*
Physics/Astronomy	0.74*	0.73*	0.68*	-0.15
Earth/Atmospheric/Marine Sciences	0.38	0.70*	0.16	-0.69*
Mathematics/Computer Sciences	0.77*	0.78*	0.67*	0.48**
Engineering	0.68*	0.74*	0.02	0.14
Agricultural Sciences	0.14	0.73*	-0.39	-0.58*
Biosciences	0.58*	0.78*	0.54*	-0.55*
Health Sciences	0.63*	0.73*	0.47**	0.42
Psychology	0.70*	0.71*	0.75*	0.39
Economics	0.44**	0.72*	-0.07	-0.55*
Social Sciences	0.67*	0.76*	0.48**	-0.47**

NOTES: (1) These are zero-order correlation coefficients. (2) Unemployment rate is lagged three years.
*Denotes correlation is statistically significant at 1% level or greater.
**Denotes correlation is significant at 5% level or greater.

TABLE 4.30: Correlation Between Percent Seeking Postgraduate Employment and the Other Market Variables

	Between SEEK and	
Field	UNEMP*	UNEMP4YR**
Chemistry	0.43	0.65
Physics/Astronomy	0.41	0.60
Earth/Atmospheric/Marine Sciences	0.55	0.71
Mathematics/Computer Sciences	0.36	0.56
Engineering	0.40	0.58
Agricultural Sciences	0.72	0.85
Biosciences	0.64	0.80
Health Sciences	0.69	0.82
Psychology	0.81	0.88
Economics	0.70	0.82
Social Sciences	0.88	0.88

*UNEMP = Overall unemployment rate for the U.S. labor force (obtained from the Bureau of Labor Statistics)
**UNEMP4YR = Unemployment rate for persons with four or more years of college (obtained from the Bureau of Labor Statistics)

TABLE 4.31: Correlation for Per Capita Number of Doctorates in the United States, 1967-1986

Field	TTD	RTD	TPGE	TNEU
Chemistry	-0.71*	-0.82*	-0.49**	0.62*
Physics/Astronomy	-0.87*	-0.93*	-0.63*	0.30
Earth/Atmospheric/Marine Sciences	-0.47*	-0.89*	-0.010	0.84*
Mathematics/Computer Sciences	-0.87*	-0.91*	-0.78*	-0.44**
Engineering	-0.83*	-0.91*	-0.16*	-0.06
Agricultural Sciences	-0.05*	-0.70	0.36	0.63*
Biosciences	-0.59*	-0.87*	-0.54**	0.70*
Health Sciences	-0.38	-0.38	-0.11	-0.39
Psychology	-0.91*	-0.89*	-0.91*	-0.68*
Economics	-0.46**	-0.91*	-0.17	0.79*
Social Sciences	-0.85*	-0.84*	-0.67*	0.36

NOTE: These are zero-order correlation coefficients.
*Denotes correlation is statistically significant at 1% level or greater.
**Denotes correlation is significant at 5% level or greater.

The Zero-Order Correlation of the Vectors

The correlations between TTD and all of the variables in each vector for which data are available for 1967-1986 are in Table 4.32. Regression analysis was used to derive an adjusted R^2 for each vector on the assumption that this is the only vector that affects TTD (no one model consistently has the highest R^2). The F vector (family background characteristics) explains most of the adjusted variation in TTD in math, health sciences, and social sciences. The I vector (individual attributes) explains most of the variation in chemistry, engineering, and psychology. Variations in two fields—agricultural sciences and biosciences—are best explained by the TLFA vector (tuition and financial aid). Finally, the O vector (organizational factors) explains most of TTD's adjusted variation in the remaining three fields: P&A; EAM; and economics. Remarkably, the E vector (economic variables) was not able to predict a larger amount of the variation than other vectors in any fields.

TABLE 4.32: Amount of Adjusted Variation in TTD Explained by Each of the Five Vectors

	Vector				
Field	F	I	TLFA	O	E
Chemistry	0.73*	0.76*	0.75*	0.71*	0.48
Physics/Astronomy	0.80*	0.77*	0.81*	0.84*	0.76*
Earth/Atmospheric/Marine Sciences	0.33**	0.29**	0.44*	0.66*	0.18
Mathematics/Computer Sciences	0.96*	0.86*	0.93*	0.95*	0.89*
Engineering	0.71*	0.85*	0.57*	0.78*	0.63*
Agricultural Sciences	0.11	0.09	0.34**	0.31	-0-
Biosciences	0.82*	0.82*	0.94*	0.84*	0.50
Health Sciences	0.91*	0.75*	0.71*	0.83*	0.41*
Psychology	0.97*	0.98*	0.94*	0.95*	0.59*
Economics	0.76*	0.64*	0.88*	0.89*	0.17
Social Sciences	0.98*	0.91*	0.95*	0.92*	0.67*

NOTE: F = Family Background (MARRIED, DEPEND, TEMP, WOMEN); I = AGE, SAMEFLD, SELECT; TLFA = TUITION, SUPFED, SUPPRIV, SUPTA, SUPRA; O = FORBACC, BTOP40, BCARN1ST, PTOP40, PCARN1ST, FACULTY, R&D; E = SALRAT1, UNEMP4YR, SEEK. Acronyms are defined in Appendix B, pp. 175-177.

*Denotes correlation statistically significant at 1% level or greater.

**Denotes correlation significant at 5% level or greater.

TABLE 4.33: Number of Fields in Which Each Variable Had a Statistically Significant Correlation with TTD or RTD

	TTD		RTD	
	Negative Correlation	Positive Correlation	Negative Correlation	Positive Correlation
	Family Background Characteristics			
MARRIED	10	0	11	0
DEPEND	10	0	11	0
BLACK	0	3	0	3
HISPANIC	0	6	0	7
WOMEN	0	10	0	11
	Student Attributes			
AGE	0	9	1	7
SELECT	3	0	4	0
SAMEFLD	2	3	3	6
	Tuition and Financial Aid			
TUITION	3	0	1	0
SUPFED	10	0	11	0
SUPPRIV	2	1	4	0
SUPRA	1	7	0	11
SUPTA	1	4	1	8
SUPOWN	0	9	0	8
	Institutional Environment			
FORBACC	2	3	2	3
FACULTY	0	8	0	11
R&D	0	9	0	11
BCARN1ST	1	2	2	1
BTOP40	3	1	4	1
PCARN1ST	3	0	6	0
PTOP40	3	0	4	0
	Salary Variables			
SDRSAL	7	3	6	4
SALRAT1	4	2	4	4
	Employment Conditions			
SEEK	0	6	0	10
DEFIN	6	0	10	0
UNEMP	0	9	0	11
UNEMP4YR	0	9	0	11
	Stock Variable			
PERPOP	10	0	9	0

NOTES: (1) These are zero-order correlation coefficients. (2) Statistical significance is at the .05 level. (3) Acronyms are defined in Appendix B, pp. 175-177.

Summary

Table 4.33 shows the number of fields with which each of the independent variables had a statistically significant correlation (p = .05) to time to the doctorate. The table is limited to zero-order correlations with TTD and its component RTD, since other components did not appear to increase TTD.

The greatest correlation to TTD was for variables indicating marital status, dependents, gender, and federal financial support. These correlations were apparent in 10 fields for TTD and in all 11 fields for RTD. As predicted by the opportunity-cost analysis, married members of the cohort and cohort members with dependents had a negative correlation to TTD. Those with federal support also showed a negative correlation to TTD, which was not predicted. Female gender was positively correlated to TTD.

Other variables that were strongly and positively correlated with RTD in all fields were research assistantships, number of full-time faculty, level of federal R&D support, the overall unemployment rate, and the unemployment rate for college graduates. The signs were not always as predicted; for example, the relation between the unemployment variables and RTD was expected to be negative but turned out positive.

Zero-order correlations must be approached with some caution. While they are useful for demonstrating an association between TTD and/or its component parts and the variables posited by the literature and by opportunity-cost hypotheses, the nature of the relationship is speculative. In addition, some of the independent variables are time-dependent and may move up or down together over long periods.

5
CHANGES IN TTD

How well does the time-series model discussed in Chapter 3 explain changes in TTD during the 1967-1986 period? To answer this question, two models are used, one based on a set of variables common to the 11 fields and a second based on a larger set of unique variables statistically significant at .05 confidence level. Although not exhaustive, the models nonetheless provide insights into what determines change in TTD. The goal of this inquiry is to answer two questions: (1) Is a unique variable or set of variables responsible for increases in TTD in the 11 fields? and (2) Is there one model that explains the change in TTD in all fields, or are the determinants of TTD specific to each field? Two different estimation models are employed to answer these questions.

Common Variables Model

Estimates derived from the common variables model are achieved in both linear and log linear form using ordinary least-squares regression. Regression results are presented in Appendix Tables 5 and 5A. A summary of the findings appears in Table 5.1. An F test indicates that all of the estimating equations are statistically significant except for agricultural sciences.[14] Differences do exist in the amount of variation in TTD explained by the equations, the standard error of the estimates, and the number of statistically significant independent variables. In six fields (chemistry, math, engineering, biosciences, psychology, and social sciences), the model explained 90 percent or more of the variation in TTD. The lowest standard errors of the estimate were found in chemistry and psychology.

---

[14] Note that the linear time-trend model in Chapter 1 suggests the absence of a trend in this field.

TABLE 5.1: Summary of Common Linear Model Regression Results for TTD, by Variable

Variable	Field(s)	Statistically Significant	Correlation (+/-)
Female	Social Sciences	yes	+
Age	Chemistry	yes	+
	Mathematics	yes	+
	Biosciences	yes	+
	Health Sciences	yes	+
	Psychology	yes	+
	Social Sciences	yes	+
Federal Support		no	
Teaching Assistantship	Psychology	yes	-
Research Assistantship	Earth, Atmospheric, & Marine Sciences	yes	-
	Psychology	yes	+
Baccalaureate from Foreign Institution		no	
Baccalaureate from Category I Research School	Chemistry	yes	-
	Psychology	yes	-
Graduate Degree from Category I Research School		no	
Number of Faculty		no	
Salary Ratio: New Ph.D.s to Ph.D.s 10 yrs after Degree	Chemistry	yes	+
	Earth, Atmospheric, & Marine Sciences	yes	-
Unemployment Rate of College-Educated	Chemistry	yes	-
Per-Capita Doctorates in United States	Chemistry	yes	-
	Engineering	yes	-
	Biosciences	yes	-
	Psychology	yes	-

Log Linear Equations

A summary of the results from the log linear equations appears in Table 5.2. In the log linear equations, the adjusted R^2s are above 90 percent in six fields, and the transformed standard errors are lower in every field than in the linear model. Further comparison of the linear and log linear estimates suggests the statistical significance of certain variables is sensitive to the model used. The log linear model does not appear to give the best estimates. Most important, a common set of variables is not responsible for changes in TTD in the 11 fields.

Weaknesses of the Common Variables Model

The common variables model has at least two important weaknesses. First, it constrains the variable set to be identical across fields even when some variables are not statistically significant. Second, many variables are included in the model, and the effects of some of the variables may be obscured by their correlation with others.

Unique Variables Model

In this model, the number of variables is varied, and additional (but not exhaustive) variables beyond those used in the common variables model are introduced. Regression analysis is used to determine which variables in each field make a statistically significant contribution to TTD. Table 5.3 (pp. 74-75) summarizes the findings obtained using this approach by field.

Summary of Findings

A summary of the regression analyses is contained in Table 5.4 (p. 76). The variable indicating female gender is significant and positive in one field in each of the three models. With the exception of age, no other variable is statistically significant in a majority of fields, although a majority of the variables are statistically significant in a limited number of fields.

Many of the variables are not robust with respect to changes in the specification of the model. For example, the sign of the regression coefficient changed for the financial aid variables as the model specification changed. Finally, the analyses indicate individual field analysis is likely to be more productive than the simple dummy-variable approach employed by Abedi and

TABLE 5.2: Summary of Common Log-Linear Model Regression Results for TTD, by Variable

Variable	Field(s)	Statistically Significant	Correlation (+/-)
Female	Biosciences	yes	+
Age	Chemistry	yes	+
	Physics & Astronomy	yes	+
	Mathematics	yes	+
	Biosciences	yes	+
	Health Sciences	yes	+
	Psychology	yes	+
	Social Sciences	yes	+
Federal Support		no	
Teaching Assistantship	Psychology	yes	-
Research Assistantship		no	
Baccalaureate from Foreign Institution		no	
Baccalaureate from Category I Research School	Chemistry	yes	-
Graduate Degree from Category I Research School		no	
Number of Faculty	Chemistry	yes	-
	Biosciences	yes	-
Salary Ratio: New Ph.D.s to Ph.D.s 10 yrs after Degree		no	
Unemployment Rate of College-Educated		no	
Per-Capita Doctorates in United States	Physics & Astronomy	yes	-
	Earth, Atmospheric, & Marine Sciences	yes	-
	Biosciences	yes	-

Benkin (1987). Each field has a set of unique variables that help explain much of the change in TTD.

Limitations of the Analysis

Because time-series analysis was used, a number of variables were highly collinear. But time and resource constraints did not permit an approach designed to isolate the unique effects of the variables. In addition, aggregation of the data to the cohort level may have obscured some of the variation within the cohorts—that is, variables affecting student decisions at the individual level may not show up as important at the cohort level. Finally, there is a problem with interpreting the age variable. While age appears to be significant in a majority of fields, the analysis does not distinguish between physiological effects and cohort effects. The possibility cannot be ruled out that age is important because it serves as a proxy for other changes experienced by the cohort. Also, older people automatically have higher TPGE.

Caution also must be taken when drawing conclusions from an analysis that relies solely on TTD. TTD is a complex quantity, the sum of many separate decisions made at different points in time. Each decision point is of interest, and there is no guarantee that the same variables impact on decisionmaking at each point. This raises the possibility that a given variable may affect decisionmaking at more than one point in a student's career. Existing literature does not provide adequate understanding of this process, and studies of the type described in Chapter 2 do not provide the insights necessary to identify the time at which individual variables impact on TTD. Additional work is needed on the lag structure implied by the model in Chapter 3 if a full understanding of the role of the independent variables is to be achieved.

Despite these drawbacks, there is a need to model TTD if only because policymakers want to understand the supply of science and engineering personnel for the labor market. A better view of the impact of the independent variables likely will be obtained using the RTD model, since the decision points at which institutional and financial variables impact are easier to pinpoint.

Finally, it should be noted that as an endpoint, TTD may be less useful in answering some questions than RTD. If the goal is to determine whether financial aid causes students to remain in graduate school longer, RTD may provide a more accurate picture of student responsiveness. Likewise, if the goal is to examine the impact of institutional environment, RTD is the better variable. However, if the goal is to understand the role of market forces, TTD may be the better choice.

TABLE 5.3: Summary of Unique Variables Model Regression Results for TTD,

Field	Variables	Correlation (+/-)	Comment
Chemistry	Age Dependents Teaching Asst. Baccalaureate from Category I Research School	+ + + -	The four variables together accounted for 92 percent of the variation in TTD. A one-year increase in age at start of doctorate increased TTD by 3.5 years. A 10 percent rise in students with baccalaureates from Category I schools reduced TTD by almost five months.
Physics and Astronomy	Age Teaching Asst. Percent Cohort Seeking Employment	+ + +	The three variables together accounted for 90 percent of the variation in TTD. A one-year increase in age boosted TTD by 2.13 years.
Earth, Atmospheric, & Marine Sciences	Research Asst. Baccalaureate from Category I Research School Percent Population with Doctorates Female	- - - +	
Mathematics/ Computer Sciences	Age Teaching Asst. Undergraduate Degree in Same Field	+ + -	A one-year increase in age increased TTD by 4.5 years, suggesting the importance of having doctoral candidates in this field entering graduate school at a young age.
Engineering	Age Percent Population with Doctorates	+ +	A one-year increase in age lengthened TTD by 1.5 years.

by Field

Field	Variables	Correlation (+/-)	Comment
Agricultural Sciences	Age	+	A one-year increase in age increased TTD by 1.1 years.
	Fed Support (decrease)	+	
	Tuition	+	
	Salary Ratio: New Ph.D.s to Ph.D.s 10 yrs. after Degree	+	
Biological Sciences	Age	+	These three variables accounted for 91 percent of the variation in TTD. A one-year increase in age lengthened TTD by 1.9 years.
	Graduate Degree from Category I Research School	+	
	Percent Population with Doctorates	-	
Health Sciences	Age	+	A one-year jump in age increased TTD by two years.
	Baccalaureate from Foreign Institution	-	
	Percent Population with Doctorates	-	
Psychology	Marital Status	+	
	Salary Ratio: New Ph.D.s to Ph.D.s 10 yrs. after Degree	+	
	Fed Support	-	
Economics	Age	+	A one-year increase in age lengthened TTD by nearly 11 months. The four variables together accounted for 84 percent of the change in TTD.
	Baccalaureate from Category I Research School	+	
	Salary Ratio: New Ph.D.s to Ph.D.s 10 yrs. after Degree	-	
	Percent Population with Doctorates	-	
Social Sciences	Age	+	A one-year increase in age boosted TTD by 1.3 years.
	Temp. U.S. Residents Receiving Ph.D.s	+	
	Salary Ratio: New Ph.D.s to Ph.D.s 10 yrs. after Degree	-	

TABLE 5.4: Number of Fields in Which Variable Has Statistically Significant Effect on TTD

	MODEL					
	COMMON				UNIQUE	
	Linear		Log Linear			
Variable	POS	NEG	POS	NEG	POS	NEG
WOMEN	1	0	1	0	1	0
AGE	6	0	7	0	9	0
SUPFED	0	0	0	0	0	2
SUPTA	0	1	0	1	3	0
SUPRA	1	1	0	0	0	1
FORBACC	0	0	0	0	0	1
BCARN1ST	0	2	0	1	1	3
PCARN1ST	0	0	0	0	1	0
FACULTY	0	0	0	2	0	0
SALRAT1	1	1	0	1	2	0
UNEMP4YR	0	1	0	1	0	0
PERPOP	0	4	0	3	0	0
MARRIED	-	-	-	-	1	0
TEMP	-	-	-	-	1	0
DEPEND	-	-	-	-	1	
SAMEFLD	-	-	-	-	0	1
TUITION	-	-	-	-	1	0
SDRSAL10	-	-	-	-	0	3
SEEK	-	-	-	-	1	0

NOTES: (1) "Pos" indicates a positive regression coefficient. "Neg" indicates a negative regression coefficient. (2) Variables below the dotted line were not entered in the common variables models. (3) For explanation of variables, see list of acronyms (Appendix B, pp. 175-177).

In short, whether TTD or RTD is the "better" dependent variable depends on which questions the researcher wishes to answer. Those studies that employ both TTD and RTD without distinguishing between the two may be ignoring the important differences between the two variables.

What Can Be Learned from the Findings?

Despite the potential problems discussed above, this time-series analysis of TTD is encouraging in several respects. It suggests that:

1. Total time to the doctorate can be modeled and such models explain much of the variation in the data in a time-series context.
2. Age is the most consistent statistically significant variable, has a large impact on TTD, and explains the largest amount of variation in the data.
3. Variables from each of the five vectors act to determine TTD. Moreover, the number of variables found to be statistically significant in this study is substantially greater than that found by Abedi and Benkin.
4. Financial aid has an impact on TTD, but not always in the intended direction. This interesting and provocative finding clearly warrants additional study in a cross-section or pooled time-series cross-section analysis.[15]
5. At least some market variables affect TTD. Since prior studies have not established this link, it opens a new avenue of inquiry for researchers interested in the determinants of time to the doctorate. It also supports the argument that market-place changes involving high-level personnel will occur as students adjust to market conditions.

However, this analysis does not suggest that sufficiently large changes in TTD can be achieved by changing financial aid policies or the institutional factors students are exposed to. It also provides little evidence that an infusion of additional resources would offset the increase in TTD.

---

[15] Aggregations of the type used here run the risk that some of the individual variation will be averaged out. Cross-section studies are almost certain to show a stronger relationship between federal support and TTD because the most promising students are the ones most likely to receive federal support and the most likely to complete degree requirements quickly.

6
CHANGES IN REGISTERED TIME TO THE DOCTORATE, TIME PRIOR TO GRADUATE ENTRANCE, AND TIME NOT ENROLLED IN THE UNIVERSITY

This chapter uses the common and unique variables models defined in Chapter 5 to explain changes in registered time to degree (RTD) and the common variables model to explain changes in time prior to graduate entrance (TPGE) and time not enrolled at the university (TNEU). As discussed in Chapter 1, TTD and RTD have a similar time trend, and increases in RTD are largely responsible for increases in TTD.

Registered Time to the Doctorate

RTD in the Common Variables Model Using Linear and Log Linear Equations

Regression coefficients for each field, using both linear and log linear estimating equations, appear in Appendix Tables 6 and 6A. A summary of the findings for each variable in each model is given in Tables 6.1 and 6.2. As was true for TTD, a comparison of the results for the linear and log estimates suggests that the results are different depending on the model used. While the importance of certain variables such as teaching assistantships, foreign baccalaureate, and salary does not change across specifications, the role of others such as age, federal support, and unemployment are affected. In most cases, the signs of the statistically significant variables do not change, and the log linear model explains the variation in the data no better than the linear model does.

RTD in the Unique Variables Model

Table 6.3 (pp. 82-83) summarizes the results of using a unique model for each of the 11 fields. Age is no longer an important variable in all fields, and no other variable has a significant impact on RTD in every field.

TABLE 6.1: Summary of Common Linear Model Regression Results for RTD, by Field

Variable	Field(s)	Statistically Significant	+/-
Female	Social Sciences	yes	+
Age	Chemistry	yes	+
	Mathematics	yes	+
	Earth, Atmospheric, & Marine Sciences	yes	+
	Social Sciences	yes	+
Federal Support	Earth, Atmospheric, & Marine Sciences	yes	-
	Biosciences	yes	-
Teaching Assistantship	Biosciences	yes	+
Research Assistantship		no	
Baccalaureate from Foreign Institution	Social Sciences	yes	+
Baccalaureate from Category I Research School	Chemistry	yes	-
	Agricultural Sciences	yes	+
Graduate Degree from Category I Research School		no	
Number of Faculty	Earth, Atmospheric, & Marine Sciences	yes	+
	Biosciences	yes	-
Salary Ratio: New Ph.D.s to Ph.D.s 10 yrs. after Degree		no	
Unemployment Rate of College-Educated	Chemistry	yes	-
	Earth, Atmospheric, & Marine Sciences	yes	-
	Social Sciences	yes	-
Per Capita Doctorates in United States		no	

TABLE 6.2: Summary of Common Log-Linear Model Regression Results for RTD, by Field

Variable	Field(s)	Statistically Significant	+/-
Female		no	
Age	Earth, Atmospheric, & Marine Sciences	yes	+
	Biosciences	yes	+
Federal Support	Biosciences	yes	+
Teaching Assistantship	Biosciences	yes	+
Research Assistantship		no	
Baccalaureate from Foreign Institution	Social Sciences	yes	+
Baccalaureate from Category I Research School	Agricultural Sciences	yes	+
Graduate Degree from Category I Research School	Agricultural Sciences	yes	-
Number of Faculty	Earth, Atmospheric, & Marine Sciences	yes	+
	Biosciences	yes	-
Salary Ratio: New Ph.D.s to Ph.D.s 10 yrs. after Degree		no	
Unemployment Rate of College-Educated	Earth, Atmospheric, & Marine Sciences	yes	-
Per Capita Doctorates in United States	Earth, Atmospheric, & Marine Sciences	yes	-

Evaluation of the Results

A number of observations can be made about Table 6.4 (p. 84), which shows the number of fields in which a particular independent variable was statistically significant. For example, no one variable explains the widely observed increases in RTD across fields. Instead, the combinations of variables

TABLE 6.3: Summary of Unique Variables Model Regression Results for RTD,

Field	Variable(s)	Correlation (+/-)	Comment
Chemistry	Age Baccalaureate from Foreign Institution Salary Ratio: New Ph.D.s to Ph.D.s 10 yrs. after Degree	+ + -	These three variables accounted for 91 percent of the variation in RTD. A one-year increase in age boosted RTD by 1.5 years. A 1 percent increase in doctorates with degrees from foreign institutions increases RTD by about a week.
Physics and Astronomy	Marital Status Graduate Degree from Category I Research School Teaching Asst.	- - +	These three variables accounted for 91 percent of variation in RTD. A 1 percent increase in married students lowered RTD by nearly two weeks. A similar increase in percentage of students from Category I school decreased RTD by a little over two weeks.
Earth, Atmospheric, & Marine Sciences	Marital Status Baccalaureate from Category I Research School Temp. U.S. Residents Receiving Ph.D.s Baccalaureate from Top-20 School	- - + +	These four variables explained 89 percent of the variation in RTD.
Mathematics & Computer Sciences	Female Salary Ratio: New Ph.D.s to Ph.D.s 10 yrs. after Degree	+ -	The two variables explained 97 percent of the variation in RTD.
Engineering	Baccalaureate from Foreign Institution Undergraduate Degree in Same Field Definite Employment	+ - -	These three variables explained 93 percent of the variation in RTD.

by Field

Field	Variable(s)	Correlation (+/-)	Comment
Agricultural Sciences	Teaching Asst. Baccalaureate from Foreign Institution Definite Employment Salary Ratio: New Ph.D.s to Ph.D.s 10 yrs. after Degree	+ + - -	These four variables accounted for 82 percent of the variation in RTD.
Biological Sciences	Research Asst. Percent Cohort Seeking Emp.. Salary Ratio: New Ph.D.s to Ph.D.s 10 yrs. after Degree	+ + -	These three variables explained 95 percent of the variation in RTD. The Durbin-Watson statistic for this regression is in the indeterminate range.
Health Sciences	Federal Support Salary Ratio: Doctorates to Baccalaureates Temp. U.S. Residents Receiving Ph.D.s	- - +	These three variables explained 85 percent of the variation in RTD. A 1 percent rise in federal support decreased RTD by about two weeks.
Psychology	Federal Support Salary Ratio: New Ph.D.s to Ph.D.s 10 yrs. after Degree Temp. U.S. Residents Receiving Ph.D.s	- - +	These three variables accounted for 96 percent of the variation in RTD.
Economics	Private Support Baccalaureate from Foreign Institution Temp. U.S. Residents Receiving Ph.D.s	- + +	These three variables explained 95 percent of the variation in RTD. A 1 percent increase in those with baccalaureate from foreign institution lowered RTD by nearly a month.
Social Sciences	Private Support Salary Ratio: New Ph.D.s to Ph.D.s 10 yrs. after Degree Temp. U.S. Residents Receiving Ph.D.s	- - +	These three variables explained 99 percent of the variation in RTD. A 1 percent jump in private support increased RTD by about a month.

TABLE 6.4: Number of Fields in Which Variable Has Statistically Significant Effect on TTD

	MODEL					
	COMMON				UNIQUE	
	Linear		Log Linear			
Variable	POS	NEG	POS	NEG	POS	NEG
WOMEN	1	0	0	0	1	0
AGE	4	0	2	0	1	0
SUPFED	0	2	0	1	0	2
SUPTA	1	0	1	0	2	0
SUPRA	0	0	0	0	1	0
FORBACC	1	0	1	0	4	0
BCARN1ST	1	1	1	0	0	1
PCARN1ST	0	0	0	1	0	1
FACULTY	1	1	1	1	0	0
UNEMP4YR	0	3	0	1	0	0
PERPOP	0	0	0	1	0	0
MARRIED	-	-	-	-	0	2
TEMP	-	-	-	-	5	0
SAMEFLD	-	-	-	-	0	1
SUPPRIV	-	-	-	-	0	2
BTOP20	-	-	-	-	1	0
SDRSAL10	-	-	-	-	0	4
SALRAT1	-	-	-	-	0	2
SALRATIO	-	-	-	-	0	1
SEEK	-	-	-	-	1	0
DEFIN	-	-	-	-	0	2

NOTES: (1) "Pos" indicates a positive regression coefficient. "Neg" indicates a negative regression coefficient. (2) Variables below the dotted line were not entered in the common variables models. (3) Acronyms are defined in Appendix B, pp. 175-177.

with statistically significant effects on RTD vary by field. In both the linear common variables model and the unique variables model, female gender was significant and positive in just one field. In the log-linear model, gender was not significant in any field. In those equations where age is statistically significant, it tends to have a large impact on RTD, suggesting that as more older students enroll in doctoral programs, RTD will increase. However, as noted earlier, age may act as a proxy for cohort differences rather than for physiological or other effects of aging. This possibility deserves more study before conclusive statements can be made. The role of financial support in affecting RTD is mixed. In a number of fields, financial variables did not enter the equation at all and, in a few, they had a positive partial correlation, contrary to intuitive expectations. This finding suggests that the effects of financial aid are field-specific and the type of aid provided influences whether students complete the doctorate more or less rapidly. The data do not allow firm conclusions about the effects of increasing financial aid as the primary source of support. The analysis suggests that in some fields increases in the number of foreign students or in the percentage of students with foreign baccalaureates have led to increased RTD. Finally, analysis supports the belief that changes in market variables—unemployment rate, salaries, and salary ratios—affect RTD.

The results of this inquiry are best viewed as suggestive rather than conclusive. Problems of multicollinearity, aggregation, and limited data suggest the need for study of these issues in a cross-section and/or pooled time-series cross-section framework. Further research is needed to affirm the role of age, to elaborate on the role of financial aid, and to provide greater insight into the role of student ability (see Chapter 7).

Time Spent Prior to Graduate School Entrance (TPGE)

The results summarized in Table 6.5 were obtained using the linear common variables model to explain changes in TPGE (see Appendix Table 7). The implicit assumption in the use of these variables is that students have prior knowledge of how their cohort is likely to fare in terms of receiving financial aid and entering the labor market.

The R^2 for the individual field equations are lower for TPGE than for TTD or RTD and, for three fields, the equations themselves are not statistically significant. In part, this results because decisions made at the time of undergraduate graduation are more likely to be based on family background and undergraduate performance factors not contained in the model (see Chapter 2). It may also be that new variables are needed to adequately capture conditions at the

TABLE 6.5: Summary of Common Linear Model Regression Results for TPGE, by Variable

Variable	Field(s)	Statistically Significant	Correlation (+/-)
Female		no	
Age	Chemistry	yes	+
	Mathematics	yes	+
	Engineering	yes	+
	Biosciences	yes	+
	Health Sciences	yes	+
	Social Sciences	yes	+
Federal Support		no	
Teaching Assistantship	Social Sciences	yes	-
Research Assistantship	Chemistry	yes	+
Baccalaureate from Foreign Institution	Mathematics	yes	+
Baccalaureate from Category I Research School		no	
Graduate Degree from Category I Research School		no	
Number of Faculty		no	
Salary Ratio: New Ph.D.s to Ph.D.s 10 yrs. after Degree	Mathematics	yes	+
Unemployment Rate of College-Educated	Mathematics	yes	+
Percent Population with Doctorates	Mathematics	yes	-

NOTE: No variables were significant for the following fields: earth, atmospheric and marine sciences; agricultural sciences; and economics.

time the decision to enter graduate school was made. For example, the relevant financial variable may be the percentage of the prior year's entering class with financial aid and the relevant market variable may be the percentage of doctorates who found jobs in the year in which the person decided to enter graduate school. Analysis of these issues may explain why fewer variables are statistically significant in the TPGE equations than in the RTD equations. It's interesting to note that in math, biosciences, psychology, and social sciences, the equations explained better than 90 percent of the variation in the data.

As was true for the linear analysis, in the log-linear analysis (Table 6.6), the equations for earth, atmospheric, and marine sciences; agricultural sciences; and economics were not statistically significant. Also, the R^2s were generally lower on these equations than for TTD and RTD.

Several points can be made about the determinants of TPGE based on the findings in this section. First, in most of the fields, the variables that explained most of the change in TPGE were demographic and economic in nature. With rare exceptions, institutional factors did not affect the TPGE. However, in the log equations the unemployment rate and salary variables were statistically significant determinants of TPGE. Second, the financial aid variables did affect TPGE in some fields, although not always in the expected direction. TPGE in chemistry and physics and astronomy was consistently affected by financial aid. Finally, in most fields neither the percentage of women nor the percentage of students with foreign baccalaureates had a statistically significant effect on TPGE.

Time Not Enrolled in the University (TNEU)

TNEU, time the student spends away from his or her studies after registering for graduate school, is affected by such factors as illness or financial exigency, frustration with the doctoral program, and the need to take a break from dissertation work (see Appendix Table 8). Since the variables in the common variables model do not specifically address these concerns, this model is not expected to explain as much of the variation in TNEU as it did for other dependent variables. Tables 6.7 and 6.8 summarize the results from the linear and non-linear regression equations.

The analysis shows no one variable consistently explained changes in TNEU in all fields. Compared to TPGE, unemployment and salary variables do not appear to have a strong effect on TNEU. This is surprising. One would expect student decisions to leave graduate school to be more affected by market conditions. And, as with TPGE, factors such as gender and percent with foreign baccalaureates do not appear to exert a strong influence on TNEU.

TABLE 6.6: Summary of Common Log-Linear Model Regression Results for TPGE, by Variable

Variable	Field(s) Significant	Statistically (+/-)	Correlation
Female	Biosciences	yes	+
Age	Chemistry	yes	+
	Physics & Astronomy	yes	+
	Engineering	yes	+
	Biosciences	yes	+
	Health Sciences	yes	+
	Psychology	yes	+
	Social Sciences	yes	+
Federal Support		no	
Teaching Assistantship	Physics & Astronomy	yes	-
Research Assistantship	Chemistry	yes	-
Baccalaureate from Foreign Institution	Mathematics	yes	+
Baccalaureate from Category I Research School		no	
Graduate Degree from Category I Research School		no	
Number of Faculty		no	
Salary Ratio: New Ph.D.s to Ph.D.s 10 yrs. after Degree	Physics & Astronomy	yes	-
Unemployment Rate of College-Educated	Physics & Astronomy	yes	+
	Psychology	yes	+
	Mathematics	yes	-
Percent Population with Doctorates		no	

NOTE: No variables were significant for the following fields: earth, atmospheric and marine sciences; agricultural sciences; and economics.

TABLE 6.7: Summary of Common Linear Model Regression Results for TNEU, by Variable

Variable	Field(s)	Statistically Significant	Correlation (+/-)
Female		no	
Age		no	
Federal Support	Biosciences	yes	+
Teaching Assistantship		no	
Research Assistantship	Biosciences	yes	+
Baccalaureate from Foreign Institution	Biosciences	yes	+
Baccalaureate from Category I Research School	Psychology	yes	-
Graduate Degree from Category I Research School	Biosciences Psychology	yes yes	+ -
Number of Faculty		no	
Salary Ratio: New Ph.D.s to Ph.D.s 10 yrs. after Degree		no	
Unemployment Rate of College-Educated		no	
Percent Population with Doctorates	Biosciences Psychology	yes yes	- -

Summary of the Findings

The common variables model appears to be more effective for understanding changes in RTD than for interpreting changes in TPGE and TNEU. No one variable is responsible for the increase in RTD over time, although in fields in which it is statistically significant, age has a relatively large effect. Moreover, the mix of variables that affect RTD is different among fields, although all five vectors described in Chapter 3 come into play.

TABLE 6.8: Summary of Common Log-Linear Model Regression Results for TNEU, by Field

Variable	Field(s)	Statistically Significant	Correlation (+/-)
Female		no	
Age	Health Sciences	yes	+
Federal Support	Chemistry	yes	+
	Physics & Astronomy	yes	+
	Biosciences	yes	+
Teaching Assistantship		no	
Research Assistantship		no	
Baccalaureate from Foreign Institution	Biosciences	yes	+
Baccalaureate from Category I Research School		no	
Graduate Degree from Category I Research School		no	
Number of Faculty	Chemistry	yes	-
Salary Ratio: New Ph.D.s to Ph.D.s 10 yrs. after Degree		no	
Unemployment Rate of College-Educated		no	
Percent Population with Doctorates	Biosciences	yes	-

NOTE: No variables were significant for the following fields: mathematics; engineering; and agricultural sciences. Only biosciences and economics had R^2s greater than 90 percent.

The linear model suggests that age has the largest impact on RTD; the percentage of students with foreign baccalaureates and who are female also consistently increases RTD. These results are field-specific and are not generalizable to all 11 fields, however. The role of financial aid is ambiguous, and different types of aid affect RTD differently.

The models explain less of the variance in TPGE and TNEU than in TTD and RTD. In some fields, the models do not produce statistically significant results. While generalizing across fields is difficult, the equations for TPGE and TNEU have fewer statistically significant variables than those for RTD and TTD. Interestingly, market variables explain time spent prior to entering graduate school while, for the most part, they are not statistically significant in the TNEU equations.

Additional work is needed to understand the factors that cause changes in TPGE and TNEU.[16] It is likely that institutional and psychological factors beyond those captured in this common variables model affect the decision to postpone entry to graduate school and/or to delay completion of the doctorate.

---

16 Knowledge of the determinants of TPGE would be useful, since it tells us how long students take to move from undergraduate to graduate school. TNEU is important because substantial differences exist across fields and we have little understanding of the underlying reasons: it may be that market opportunities for ABDs are substantially different among fields or that some field work is useful before obtaining the doctorate.

7
PAST AS PROLOGUE

> If the preparation of college teachers and the national distribution of graduate study are the two major issues in graduate education today, then the duration of doctoral study is probably the third. The critics who fear that the system is going to turn out too few doctorates in the years ahead, those who believe that the whole emphasis on research is wrong, those who think that the degree has fallen off from traditional standards, even those who want things added—all of them are concerned about the lengthy period of doctoral study. There is hardly a recent discussion of graduate education in which this note is not played loud and strong.
>
> (Berelson, 1960:156)

What Has Happened to Time to the Doctorate?

Total Time to the Doctorate

Despite ample evidence that TTD has been increasing for years, public attention to the question of how long it should take to complete the doctorate has diminished. The extent of the change in TTD between 1960 and the present is highlighted by a comparison of Berelson's data with data from this study (Table 7.1). If current trends persist, it will take even longer for doctorates to complete their degrees in the future. This is an important conclusion because it suggests that the question of whether doctoral preparation could, or should, be expedited may again become a matter of great interest.

Unfortunately, Berelson lacked the data to study long-term changes in RTD. His study used data from only one year and focused on the difference between these two variables and TTD. It found that RTD was lower than TTD in

TABLE 7.1: Median Total Time to the Doctorate Over Time

	Berelson			Doctorate Records File		
Aggregated Field	1936	1957	1967	1977	1987	1997
Physical Sciences	6	6	5.9	6.9	7.1	7.5
Biological Sciences	6	7	6.7	6.9	8.0	8.4
Social Sciences	8	8	7.6	7.9	10.4	11.2

NOTE: The figures for 1997 are estimated using a simple time-trend model.

each of eight fields under study.[17] Of particular note, according to Berelson, was the fact that the time differences among fields were small when actual time to the doctorate was considered.[18] He concluded that "[t]he problem is not how much time a student should spend in working on his degree, but rather over how long a period of time he should do it" (Berelson, 1960:162).

Registered Time to the Doctorate

Because RTD data are available for both 1967 and 1986, it is possible to look at RTD over time. In all 11 fields, it increased, sometimes by a large amount. In seven fields, RTD increased more than TTD between 1967 and 1986. For example, RTD rose by 49 percent in the social sciences, compared to a 22-percent increase in TTD; in economics, the comparable figures were 37 percent and 4 percent; in earth, atmospheric, and marine sciences, 28 percent and 14 percent; and in agricultural sciences, 22 percent and 8 percent. In three fields, RTD and TTD increased by a similar percentage: about 28 percent in psychology; 13 percent in physics; and 29 percent in math and computer sciences. Only in the health sciences did the change in TTD (27 percent) greatly exceed the change in RTD (14 percent) between 1967 and 1986. These findings suggest that, with the exception of one field, the major source of increasing TTD was a "stretching-out" of the time spent registered in graduate school.

---

[17] These fields are physical sciences, biosciences, social sciences, humanities, engineering, education, arts and sciences, and professional fields.

[18] The lowest median actual time was in education (2.8 years) and the highest was in social sciences (3.7 years).

The differences among fields in RTD described by Berelson can be examined for more recent years using data from the DRF. For both 1967 and 1986, the difference in median RTDs across fields is less than the difference in median TTDs, affirming Berelson's findings.

The range in TTD between the high and low fields increased 1.9 years from 1967 to 1986. The lowest mean TTD in 1967 was 6.4 years (for chemistry) and the highest was 10.6 years (for the social sciences). The range between the low and high fields, therefore, was 4.2 years. In 1986 the field with the lowest mean (7.2 years) was again chemistry, but the field with the highest mean was health sciences (13.3 years). In this case the difference between the two fields was 6.1 years.

The range in RTD also grew between 1967 and 1986, but that growth was less than that experienced by TTD. In 1967, chemistry had the low mean RTD (5.0 years) and health sciences had the high mean (6.5 years). The range between the two is 1.5 years. In 1986, the low field was still chemistry with a mean RTD of 5.8 years; the high field was psychology, with a mean of 7.5 years. The difference between the two fields is 1.7 years, compared to 6.1 years using the TTD measure, and the range between high and low fields for RTD grew by 0.2 years from 1967 to 1986, far less than the 1.9 year growth observed using TTD.

Thus, although Berelson found that the RTD measure produced a smaller difference across fields, he failed to see that the range was increasing over time, suggesting the doctorate is growing relatively more costly in certain fields in terms of lost income while in graduate school.

Variation Around the Mean

To determine whether within-field differences in the time students took to earn the doctorate narrowed or grew larger between 1967 and 1986, coefficients of variation (CVs)[19] were computed for each field. The results show that the within-field variation in both RTD and TTD was at least as large as between-field variations in some fields, raising the question of whether the type of field comparisons offered by Berelson are useful.

In all 11 fields, the CVs for TTD decreased from 1967 to 1986. However, the CVs for mean RTD increased in four fields, remained the same in two, and fell in five. This indicates a larger proportion of doctorate recipients

---

[19] The coefficient variation is the standard deviation divided by the mean. It is used to express variation in the data relative to the mean and facilitates comparison of variation across fields.

had TTDs closer to mean TTD in 1986—that is, mean TTD was representative of a larger percentage of the cohort—than was the case in 1967 and a larger proportion of the 1986 than 1967 doctorate cohort took longer time to complete the doctorate (this was also true for the five fields in which the CVs for RTD fell[20]). The lengthening of time to the doctorate is affecting a larger percentage of doctorate recipients than was true in the past.

Could Changes in TPGE and TNEU Have Been Large Enough To Explain the Change in TTD?

The data suggest that time prior to entry to graduate school (TPGE) rose in all fields except EAM and agricultural sciences. The size of the increase depended on the field studied, with three fields showing an increase of less than 10 percent, in three a jump of 11-50 percent, and in three a rise of 60-105 percent. The largest increases in TPGE were in math and computer sciences (105 percent) and the health sciences (100 percent), while the smallest were in economics (5 percent) and the social sciences (8 percent). Measured in absolute terms, the increases in TPGE were fairly small. In six of the nine fields in which TPGE grew, the increase amounted to less than three months.

Three other insights emerge from a study of TPGE. First, the low TPGEs for most fields in 1986 suggest that most doctorate recipients entered graduate school soon after completing the baccalaureate. And, while TPGE rose in a majority of fields, the increase was not great enough to explain more than a small fraction of the increase in TTD between 1967 and 1986.[21] Three of the four fields with large increases in TTD also had large increases in TPGE: health sciences, math, and psychology. However, even in these fields, the rise in TPGE was not large enough to be the prime source of the increase in TTD. Third, the data also suggest that changes in TNEU were not responsible for the growth in TTD in most fields. TNEU decreased in eight fields, and in five of these the decrease was greater than three months. TNEU rose by two-and-a-half months in

---

[20] The coefficient of variation dropped by 10 percent in health sciences, by 6 percent in social sciences, by 4 percent in psychology, by 5 percent in the biosciences, and by 1 percent in chemistry.

[21] For example, the rise in TPGE represented 19 percent of the growth in chemistry, 22 percent in math, 25 percent in psychology, and 37 percent in health sciences.

math and by nearly a year in health sciences; however, only in the latter was the combined effect of changes in TPGE and TNEU large enough to have a large impact on TTD. In fact, the decline in TNEU in many fields helped to offset the relatively small rises in TPGE, causing RTD to become the major source of change in TTD. These findings suggest that the concern expressed in the 1960s over the amount of time students spend "outside the system" is not valid at present (Wilson, 1965).

Possible Explanations

Six broadly based theories may explain the growth in TTD. These categories of explanation correspond to, but encompass more than, the vectors used in the model introduced in Chapter 3. The theories—Epistemic, Institutional, Student Preference-Based, Financial Need-Based, Demographic and Ability-Based, and Market-Based—are not mutually exclusive but provide a useful way of classifying the arguments made in earlier studies to justify increases in TTD.

Epistemic Explanations

The underlying premise of these explanations is that an expanding knowledge base requires that students take more time to learn, absorb, and retain what is needed to earn the doctorate. A corollary is that more (and perhaps higher quality) work is expected of the doctoral student now than in the past. But measurement of an epistemic trend requires an objective measure of the expansion of knowledge in each field. While indirect indices of this expansion (such as counts of pages, books, journals, courses, and citations) are available, there is no consensus on how to define the body of thought a doctoral student must master. Similarly, it is difficult to agree on the length of time a student should be given to master the body of knowledge required for a doctorate, since students progress at different rates. To limit the time needed to earn the doctorate is to run the risk of excluding potentially productive scholars. More research is needed to pinpoint changes in the prerequisites for entry to the graduate program, in course load, and in the standards used to judge a dissertation within each field.

Institutional Explanations

Factors in the university and/or departmental environment—such as goals and commitment, the interaction between faculty and students, and changes in student attitudes toward themselves and their peers—can also affect TTD. This study indirectly measures changes in the institutional environment over time by looking at the quality of the doctoral department, the type of undergraduate and graduate institution attended, and the effects of changes in selected resources. These aggregate measures are not substitutes for the more specific sociological and institutional variables described by Wilson (1965).

The analysis indicates that changes in the percentage of a cohort at a top-ranked graduate department do not affect either TTD or RTD. Interestingly, however, increases in the percentage of a cohort whose baccalaureate was earned at a first-tier doctorate-granting university do reduce TTD and RTD, albeit in a limited number of fields; but there is no evidence that a graduate department's high quality rating is associated with a low mean TTD.

The analysis also fails to establish a link between aggregate resource intensity, such as the aggregate number of faculty and R&D spending, and TTD. We cannot rule out the possibility that such evidence would have been found if the data series for these variables had been field-specific. Given the gross measures used and the limited number of observations available, our findings for these variables should be viewed as suggestive rather than conclusive.

Clearly, additional work is needed to flesh out the impact on RTD of the institutional environment. At present, it is not clear whether RTD is increasing because students are taking more courses, because they spend more time working while registered at the university, because more prerequisite courses are required, or because it simply takes longer to complete the dissertation. Additional work also is needed to develop causal models of institutional factors. Such studies might merge institutional and departmental data with data on average student performance and progress within the department over several years.

Student Preference-Based Explanations

This explanation assumes students prefer to stretch out their graduate training because they like being "perennial students," graduate school offers a desirable environment, students prefer to allocate time in graduate school to nondoctorate-related activities, and/or they fear they won't be able to find a job after graduation. These preferences are not easily captured in a time-series model because no consensus exists on which student attitudes should be measured and on how to measure them and, at present, the Survey of Earned Doctorates, the only yearly study of doctoral students, does not collect information on graduate student preferences over time.

Many factors can cause students to change their reasons for attending or for leaving graduate school. Decisions by university administrators may make the graduate school environment less comfortable or may place limits on financial aid. And societal mores may put pressure on those who remain outside the labor force too long. In addition, students also may change their perceptions of the benefits of a college education. Clearly, these factors can alter both RTD and TTD.

This study introduces student choice into a time-series model by looking at behavior at the margin. Of primary concern is whether changes in the marketplace cause students to alter their choices regarding graduate school.

Financial Need-Based Explanations

The financial pressures on students may increase as a result of illness or injury, tuition increases, marriage, family obligations, reduced loan or financial aid packages, and/or increases in the cost of living. Because of these factors, students may find it necessary to spend more time working and less time studying, thereby increasing TTD through effects on TPGE, RTD, and TNEU. Marital status and increases in family size raise TTD in a few fields but do not provide a general explanation of why TTD has increased in all 11 fields in this study. Changes in students' domestic situations contribute to the rise in TTD but are not the primary cause.

An argument can also be made that TTD and RTD may have risen because fewer students are receiving federal financial aid. Wilson's study found that the percentage of those with financial aid was greater among those students who finished the doctorate quickly than among those who took more time to finish. It also reported that about one-third of the students who delayed entry to graduate school did so for financial reasons. This, among other things, led Wilson to recommend increases in financial aid as a way to hasten TTD. While Wilson's evidence is suggestive, it poses a problem of causality. Did students who are recipients of financial aid finish faster because they had such aid or because such aid was given to the most able? This question remains to be answered. Moreover, Wilson's study ignored the question of whether the form of financial aid made a difference for TTD and made no attempt to quantify the effects of financial aid on the several times to the doctorate.

A comparison of the mean TTDs of those receiving federal fellowships, TAs, RAs, and private foundation support to the mean TTDs of those whose primary source of support was their own earnings (Table 3.1, p. 40) revealed that those who provided their own financial support took substantially longer to complete the doctorate than those with other types of support. Interestingly, mean TTD either fell or stayed constant between 1986 and 1987 for TA holders in seven fields and for federal fellowship holders in eight fields; it rose in seven fields for RA holders and for those who provided their own support.

The effect of financial aid on TTD is not as apparent in the causal models presented in Chapters 5 and 6. This is, in part, because the variables used in the model do not focus on the primary source of support. Moreover, the role of the financial aid variables may be obscured by their correlation with other independent variables in the model. The findings suggest that when it is a significant factor, it has a limited effect on TTD (relative to the total time required to complete the doctorate) and RTD. For example, using the linear common variables model, a 10-percent increase in the number of psychology students with TAs results in a decline of just four months in TTD. In fact, none of the financial aid variables had a consistent effect on TTD and, in some fields, they did not change TTD at all.

In the TPGE equations, federal support was not statistically significant in any field; TA support had a negative effect in one field; and RA support had a positive effect in one field. In the TNEU equations, federal support had a positive effect in one field; TA support was not statistically significant; and RA support was positive in one field.

Recent DRF surveys have collected new data on prime source of financial aid. These data could be used to analyze more thoroughly the effect of financial aid on the four dependent variables.

Demographic and Ability-Based Explanations

In recent years, doctoral students are more likely to be older, female, foreign, and minority, all factors that can increase TTD and RTD. Recent interest in certain demographic factors probably is a response to trends in the DRF data. For example, in 1976 women constituted just 22 percent of the 18,583 science and engineering doctorate recipients. By 1985, women represented 27 percent of the 19,164 science and engineering doctorate recipients (Coyle, 1986). Likewise, the share of non-U.S. citizens with permanent or temporary visas who received science and engineering doctorates grew from 21 percent in 1976 to 27 percent in 1985. Given the changing composition of the doctorate-recipient group, a natural question arises as to whether these changes were responsible for the increase in TTD.

Gender, residency status, and race do not consistently affect the measures of time to the doctorate in the 11 fields studied. In fact, the only demographic variable that has a large enough effect across fields to affect TTD is age at entry to graduate school. Age is important in the TTD, RTD, and TPGE equations but does not have a statistically significant effect in most fields in the TNEU equations. Unfortunately, the analysis does not distinguish whether age is a proxy measure or truly reflects the effects of aging on learning.[22]

---

[22] We cannot dismiss the possibility that changes in student abilities were a major factor. The lack of student skills data, such as GRE scores, did not allow study of this possibility, however.

Market-Based Explanations

Employment opportunities, the absolute salaries of doctorate holders, relative salaries, and the rate of return to alternative careers all affect time to the doctorate. Their impact is felt both by those in graduate school and by those considering alternative fields of graduate study. The assumption is that when the economic return for graduating with a doctorate falls relative to the return to nondoctorates, TTD will rise. Economic return diminishes in a given field if the unemployment rate of new doctorates rises relative to those without a Ph.D., if the relative salary of nondoctorates rises relative to that of new doctorates, and if the earnings of Ph.D.s fail to progress as rapidly over time as the earnings of those without doctorates. The longer a student remains in graduate school, the less economic return is expected.

The results of this study suggest that changes in the marketplace were not large enough or pervasive enough to provide the primary explanation for the observed increases in TTD. Increases in the unemployment rate for those with four or more years of college education reduced RTD in four fields in one model while increased unemployment affected TTD in only one field. Changes in the percentage of students seeking employment and of those with definite postgraduate plans affected TTD and RTD, but only in a few fields. TTD fell in some fields as salaries for experienced doctorates rose, and it increased when there was a decline in the salary of new doctorates relative to salaries of doctorates 10 years postgraduation. Additional modeling is needed to confirm these findings and to identify the appropriate lags between market changes and changes in TTD.

Is There A Single Explanation for Increase in TTD?

A series of factors, rather than one explanation, appears to be responsible for the trend of increasing TTD across fields. Part of the increase in TTD probably was due to epistemic factors, but this theory does not explain why there was three times the growth in TTD in the social sciences compared to chemistry (nine months versus 2.4 years). It seems unlikely that growth in the knowledge base alone could explain such a large increase in TTD in one field and a relatively small increase in another.

Institutional factors also came into play. Likewise, declining enrollments in some institutions may have created an incentive for them to keep students longer. Although the institutional environment may not have been stable during the period of study, it is not clear that these factors explain the inter-field changes described.

Among demographic variables, age is important because older students wait longer to enter graduate school and also spend more time registered in graduate school than younger students. The finding that older students take longer to complete the doctorate warrants further study. In some fields, variables such as residency and gender also affected TTD, as did financial need. This study also suggests that market forces, particularly increases in the unemployment rate and in the salaries of doctorates and nondoctorates, affect TTD.

The finding that no one class of explanations is responsible for the rise in TTD is consistent with the initial correlations in Chapter 4 and with the set of regressions presented in Chapters 5 and 6. It also confirms Wilson's 1965 findings of the multi-factorial aspect of any steps taken to reduce TTD:

> In essence, the amount of time involved in doctoral preparation can be reduced, our respondents indicate, only through concerted effort on a variety of fronts. Solutions predicated on a monistic conception of the problem will not prove to be satisfactory and no approach to "time reduction" stressing only one line of attack, e.g., increased financial support, . . . will be sufficient, however necessary it may be to an overall solution.

As has been shown, TTD is affected by a number of variables. But aggregate models alone cannot identify steps to reduce TTD. What is needed is a more disaggregated study of what is happening at the department level. And additional modeling should be done using the student as the unit of analysis to sort out the roles of ability level, past preparation, and financial aid in elongating TTD. Existing studies do not provide sufficient guidance for policymakers to reduce TTD.

Implications of a Continuing Rise in TTD

A More Resource-Intensive Doctoral Program

Changes in TTD that result from an increase in time spent in graduate school will raise the cost (excluding opportunity costs) of obtaining a doctoral degree. The annual cost, on average, to educate a graduate student ranges between $21,855 and $29,235. The mid-range estimate is $25,545 per year.[23]

---

23 The U.S. Department of Education, National Center for Education Statistics (NCES), *Digest of Education Statistics 1988*, indicates that educational and general expenditures per FTE university student were $13,179 in 1985-86 (Table 243). We have assigned weights to account for the higher cost of graduate

The fields with the smallest rise in RTD between 1967 and 1986 (0.8 years) were engineering, chemistry, and physics and astronomy; the field with the largest increase (2.9 years) was the social sciences. Using 1967 as the base year, the percentage increase in RTD was 14 percent in engineering and 49 percent in the social sciences. Assuming the cost of programs does not vary across fields, the cost of a doctorate rose by $20,436 ($25,545 x 0.8) in engineering and by $74,081 ($25,545 x 2.9) in the social sciences between 1967 and 1986. Taking all graduates into account, the increase in RTD caused an additional $35,190,792 ($20,436 x 1,722) to be spent educating engineering doctorate recipients and an additional $106,602,550 ($74,081 x 1,439) in outlays to educate social science doctorate recipients.

Graduate students themselves pay only a small fraction of the $25,545 average yearly cost of graduate training, with the rest coming from other sources.

A Longer Gestation Period

Increases in TTD force employers to wait longer to hire new doctorates, potentially causing a shortage of trained workers in affected fields and driving up the salaries of those who already hold doctorates. Lengthening TTD can also contribute to a public perception of shortage and thereby increase pressures for public subsidies in fields in which trained doctorates appear to be in short supply. Increases in TTD may also cause increased demand for foreign-trained doctorates.

Lengthening TTD also means the productive output of doctorates will fall. For example, suppose the average age of graduate students in the social sciences at time of entry to graduate school was 27 years in 1967. If RTD in 1967 was 6 years, the average doctorate holder would graduate at age 33. If that person had no periods of unemployment and utilized his or her doctorate knowledge until retirement at age 65, a total of 32 person-years of work would have been produced. But if, in 1986, the average RTD rose to 9 years, the new doctorate's entry into the labor force would be delayed until age 36, reducing the average number of productive person-years to 29, a decline of 9.4 percent. If

---

education: weight 1 for part-timers and weight 2 or 3 for full-timers. NCES estimates that 56 percent of doctoral students were full-timers in 1986-87. Thus, the range of expenditures is $20,559 to $27,939, with a midpoint of $24,249. To these institutional costs are added the students' costs of doctoral education, estimated at $2,874, derived from NCES' National Postsecondary Student Aid Study, which found a cost of $3,126 for full-timers and $2,554 for part-timers.

TPGE also increased during the period, the number of productive person-years of effort would decline even more.[24]

Clearly, increases in RTD and TPGE may reduce the productive worklife of a new doctorate and reduce the overall number of high-level personnel available to employers. More doctorates would have had to be produced in 1986 than in 1967 to obtain the same number of person-years of work as in 1967. In fact, however, there was no increase in the number of new doctorates; DRF data indicate the number of new doctorate recipients has remained relatively constant since 1970 (Coyle, 1986). Although work yield of a given cohort of new doctorates is affected by a variety of factors, including mobility patterns, obsolescence, and economic conditions, this simple example illustrates that changes in TTD can affect labor supply.

Longer TTDs also slow job market response to increases in demand. There is normally a lag when engineering and scientific labor markets adjust to changes in demand (Tuckman, 1988). As the length of time required to produce a doctorate increases, so too does the length of time needed for supply to respond to increases in demand. And sudden increases in demand were more likely in 1986 than in 1967 to cause a longer period of market disequilibrium. The long-term effect of an increase in TTD is to reduce the responsiveness of high-level labor markets.

Increased Attrition

To the extent that increases in RTD are due to factors beyond student control—such as increased financial pressures, frustration created by the length of time required to complete the doctorate, of "better" opportunities—some students may choose to abandon their graduate studies altogether. The literature review uncovered no studies that looked at how changes in RTD and TTD affected student attrition, but it seems likely that, at the margin, some students consider cost when deciding to forego an additional year of graduate school. To the extent that this phenomenon occurs, increases in TTD will reduce the number of people who complete the doctorate. Such attrition will also increase the costs to society of producing a trained doctorate.

---

24 This analysis assumes no change in retirement behavior. The effect of lengthening TTD on productivity will not be as dramatic if retirement age is rising.

Lower Returns for Graduate Study

Longer TTD increases the costs of doctoral education. Even students with fellowships incur an opportunity cost because this type of support is less than the earnings that they would have received in a full-time job. Also, as noted, increases in RTD reduce the number of productive years during which a student can realize a return on his or her investment.

To the extent that students view graduate study as a potential investment, reductions in return from doctoral education are also likely to affect the decision whether to obtain a degree at all. Some students may find changes in TTD have made alternatives to a doctoral degree more attractive. For example, in many graduate schools, the Master's of Business Administration degree takes only two years to complete; thus, if the TTD required to obtain a doctorate in the sciences increases, some students will opt instead to obtain an MBA. A similar phenomenon may occur as students consider an advanced degree in medicine, law, or other professional fields. To the extent TTD rises less slowly in these fields, the relative return for obtaining a degree in them increases. Over time, more students may be drawn away from fields with high TTDs and into fields with shorter TTDs, leading to a possible shortage of trained scientists and engineers in certain high-TTD fields.

Changes in the Attractiveness of Alternative Doctorate Careers

Students choose a major based on expected returns (Chapter 2)—that is, the earnings they can expect to receive after earning the degree. Differences exist in the rate at which TTD and RTD are growing among fields. Thus, a person with an undergraduate degree and an interest in one field—physics, for example—may nonetheless choose advanced study in another field—perhaps mathematics—because the expected returns to a doctorate in the latter field are higher. To the extent that this occurs, a shortage may eventually develop in those fields with relatively larger TTDs.

TTD as a Policy Instrument

This study was motivated by interest in manipulating TTD to meet possible difficulties in producing a future supply of doctorates that will be adequate to meet anticipated needs. It is easy to argue that the increase in TTD can be reversed by increasing the number of federal fellowships or by granting more teaching and research assistantships, but the findings of this report suggest we need to learn more about the effects of the various types of financial aid before assessments of the desirability of such a solution can be made. Data are simply not available to permit policymakers to choose the best way to affect TTD or to assess the consequences of the various alternative solutions proposed by other studies.

SELECTED BIBLIOGRAPHY

Abedi, J., and E. Benkin. 1987. The effects of students' academic, financial, and demographic variables on time to the doctorate. Research in Higher Education 27(1):3-14.

Alexander, E., and D. Frey. 1984. An econometric estimate of the demand for M.B.A. enrollment. Economics of Education Review 3:97-103.

Alwin, D. F. 1974. College effects on educational and occupational attainments.American Sociological Review 39:210-223.

Astin, A. 1962. Productivity of undergraduate institutions. Science 136:129-135.

_____, and R. Panos. 1968. The Educational and Vocational Development of College Students. Washington, D.C.: American Council on Education.

Baird, L. L. 1976. Who goes to graduate school and how they get there. In Scholars in the Making, J. Katz and R. T. Hartnett, eds. Cambridge, Mass.: Ballinger, pp. 19-48.

Bean, J. P. 1980. Dropouts and turnover: The synthesis and test of a causal model of student attrition. Research in Higher Education 12:155-187.

_____. 1985. Interaction effects based on class level in an explanatory model of college student dropout syndrome. American Educational Research Journal 22:35-64.

Becker, W. 1986. The Demand for Higher Education. Paper prepared for the National Science Foundation, Division of Policy Research and Analysis.

Belliveau, J., E. Kealey, and M. Von Zur-Muehlen 1981. Doctoral enrollment and graduation patterns at Canadian universities during the seventies and their implications for the eighties: A statistical documentation by discipline. ERIC Document No. 203.795.

Benkin, E. M. 1984. Where Have All the Doctoral Students Gone? A Study of Doctoral Student Attrition at UCLA. Ph.D. dissertation. University of California at Los Angeles.

Berelson, B. 1960. Graduate Education in the United States. New York: McGraw-Hill.

Berg, H. M., and M. A. Ferber. 1983. Men and women graduate students: Who succeeds and why? Journal of Higher Education 54:629-648.

Berger, M. C. 1988. Predicted future earnings and choice of college major. Industrial and Labor Relations Review 41(3):418-429.

Berryman, S. 1982. The Adjustments of Youth and Educational Institutions to Technologically-Generated Changes in Skill Requirements (Research Report No. 85-08). Washington, D.C.: National Commission for Employment Policy.

Biglan, A. 1973. Relationships between subject matter characteristics and the structure and output of university departments. Journal of Applied Psychology 57(3):204-213.

Bowen, H. R. 1953. Graduate Education in Economics, Part 2 (Supplement). American Economic Review 43(4).

Brennan, M. A. 1985. Where have all the scholars gone...long time passing....ERIC Document No. 263.045.

Campbell, R., and B. Siegel. 1967. The demand for higher education in the United States, 1919-1964. American Economic Review 57:482-494.

Carmichael, O. C. 1961. Graduate Education: A Critique and a Program. New York: Harper and Brothers.

Carroll, S., B. Mori, D. Relles, and D. Weinschrott. 1977. The Enrollment Effects of Federal Student Aid Policies. Santa Monica, CA: Rand Corporation.

Cebula, R. J., and J. Lopes. 1982. Determinants of student choice of undergraduate major field. American Educational Research Journal 19(2):303-312.

Centra, J., and D. Rock. 1969. College environments and student achievement. American Educational Research Journal 8:305-316.

Chase, J. L. 1964. The numbers game in graduate education. Journal of Higher Education. 35:138-141.

Coyle, S. 1986. Summary Report 1985: Doctorate Recipients from U.S. Universities. Washington, D.C.: National Academy Press.

_____. 1987. Summary Report 1986: Doctorate Recipients from U.S. Universities. Washington, D.C.: National Academy Press.

Creswell, J. W., and J. P. Bean. 1981. Research output, socialization, and the Biglan model. Research in Higher Education 15:69-91.

Davis, J., with D. Gottlieb, J. Haida, C. Huson, and J. Spaeth. 1962. Stipends and Spouses: The Finances of American Arts and Science Graduate Students. Chicago: University of Chicago Press.

Davis, J. A. 1966. The campus as a frog pond: An application of the theory of relative deprivation to career decisions of college men. American Journal of Sociology 72:17-31.

Dolph, R. F. 1983. Factors Relating to Success or Failure in Obtaining the Doctorate. Ph.D. dissertation. Georgia State University.

Ethington, C. A., and J. C. Smart. 1986. Persistence to graduate education. Research in Higher Education 24(3):287-303.

Ferber, M. A., and W. McMahon. 1979. Women's expected earnings and their investment in higher education. Journal of Human Resources 14:405-420.

Freeman, R. 1971. The Market for College Trained Manpower. Cambridge, Mass.: Harvard University Press.

Freiden, A., and R. Staaf. 1973. Scholastic choice: An economic model of student behavior. Journal of Human Resources 8:396-404.

Girves, J. E., and V. Wemmerus. 1988. Developing models of graduate student degree progress. Journal of Higher Education 59:163-189.

Gropper, L., and R. Fitzpatrick. 1959. Who Goes to Graduate School? Pittsburgh: American Institute for Research.

Hansen, W. L., H. Newburger, F. Schroeder, D. Stapleton, and D. Youngday. 1980. Forecasting the market for new Ph.D. economists. American Economic Review 70:49-64.

Hartnett, R. T. 1985. Trends in Students in Doctoral and Professional Education. New Brunswick, N. J.: Rutgers University.

Hearn, J. C. 1987. Impacts of undergraduate experiences on aspirations and plans for graduate and professional education. Research in Higher Education 27(2):119-141.

Heath, J. A., and H. P. Tuckman. 1986. The effects of tuition level and financial aid on the demand for advanced terminal degrees. Economics of Education Review 6(3):227-238.

_____. 1989. The impact on labor markets of the relative growth of new female doctorates. Journal of Higher Education 60(6):704-715.

Heyns, B. 1974. Social selection and stratification within schools. American Journal of Sociology 79:1434-1451.

Jacks, P., D. Chubin, A. Porter, and T. Connolly. 1983. The ABCs of ABDs: A study of incomplete doctorates. Improving College and University Teaching 31:74-81.

Jackson, G., and G. Weathersby. 1975. Individual demand for higher education: A review and analysis of recent empirical studies. Journal of Higher Education 46:623-652.

Knapp, R. H., and H. B. Goodrich. 1952. Origins of American Scientists. Chicago: University of Chicago Press.

_____, and J. J. Greenbaum. 1953. The Younger American Scholar: His Collegiate Origins. Chicago: University of Chicago Press.

Koch, J. V. 1972. Student choice of undergraduate major field of study and private internal rates of return. Industrial and Labor Relations Review 26:680-685.

Kolman, E. M., K. S. Gallagher, D. Hossler, and F. Catania. 1987. The outcomes of doctoral education: An institutional study. Research in Higher Education 27(2):107-118.

Leslie, L., and P. Brinkman. 1986. Student Price Response in Higher Education: The Student Demand Studies. Center for the Study of Higher Education, College of Education, University of Arizona.

McDonough, J. E., and G. F. Wagstaff. 1983. Occupational perceptions of academic disciplines. Journal of Vocational Behavior 23:251-256.

Pascarella, E. T. 1980. Student-faculty informal contact and college outcomes. Review of Educational Research 50(4):545-595.

_____, ed. 1982. Studying Student Attrition: New Directions for Institutional Research, Vol. 36. San Francisco: Jossey-Bass.
_____. 1984. College environmental influences on students' educational aspirations. Journal of Higher Education 55(6):751-771.
_____, and P. T. Terenzini. 1979. Student-faculty informal contact and college persistence: A further investigation. Journal of Educational Research 72:214-218.
_____, and P. T. Terenzini. 1983. Predicting voluntary freshman year persistence/withdrawal behavior in a residential university: A path analytic validation of Tinto's model. Journal of Educational Psychology 75(2):215-226.
_____, P. T. Terenzini, and J. Hibel. 1978. Student-faculty interactional settings and their relationship to predicted academic performance. Journal of Higher Education 49(5):450-463.
Pressey S. L. 1962. Age and the Doctorate--Then and Now. Journal of Higher Education. 33:283-287.
Prior, M. E. 1962. A manifesto on graduate education. Journal of Higher Education 33(5):283-287.
Rossi, A. S., and A. Calderwood. 1973. Academic Women on the Move. New York: Russell Sage Foundation.
Smart, J. C., and E. T. Pascarella. 1986. Socioeconomic achievements of former college students. Journal of Higher Education 57(5):529-549.
Solmon, L. C. 1981. New findings on the links between college education and work. Higher Education 10:615-648.
Spady, W. 1970. Dropouts from higher education: An interdisciplinary review and synthesis. Interchange 1:64-85.
Spaeth, J. 1968. The allocation of college graduates to graduate and professional schools. Sociology of Education 41(4):342-349.
Summerskill, J. 1962. Dropouts from college. In The American College: A Psychological and Social Interpretation of the Higher Learning, Nevitt Sanford, ed. New York: John Wiley & Sons, Inc., pp. 627-657.
Teague-Rice, L. 1981. A Profile of the Female Doctoral Student Who Persisted to the Completion of the Doctoral Degree. Ph.D. dissertation. Auburn University.
Thistlethwaite, D. L. 1959. College environment and the development of talent. Science 130:71-76.
_____. 1962a. College press and student achievement. Journal of Educational Psychology 50:183-191.
_____. 1962b. Fields of study and development of motivation to seek advanced training. Journal of Educational Psychology 53(2):53-64.
Tierney, M. 1980. The impact of financial aid on student demand for public/private higher education. Journal of Higher Education 51:527-545.
Tinto, V. 1975. Dropout from higher education: A theoretical synthesis of recent research. Review of Educational Research 45(1):89-125.

Trusheim, D., and J. Crouse. 1981. Effects of college prestige on men's occupational status and earnings. Research in Higher Education 14(4):283-304.

Tucker, A., D. Gottleib, and J. Pease. 1964. Attrition of Graduate Students at the Ph.D. Level in the Traditional Arts and Sciences (Publication No. 8). East Lansing, Mich: Michigan State University Office of Research Development and the Graduate School.

Tuckman, H. 1988. Supply, human capital, and the average quality of the labor force. Economics of Education Review 7(4).

Wallace, W. L. 1965. Peer influence and undergraduate aspirations for graduate study. Sociology of Education 38:375-392.

Weidman, J. C. 1984. Impacts of campus experiences and parental socialization on undergraduates' career choices. Research in Higher Education 20:445-476.

Wilson, K. L. 1978. Toward an improved explanation of income attainment: Recalibrating education and occupation. American Journal of Sociology 84:684-697.

Wilson, K. M. 1965. Of Time and the Doctorate. Atlanta: Southern Regional Education Board.

Wilson, R., J. Gaff, E. Diesnt, L. Wood, and J. Bavry. 1975. College Professors and Their Impact on Students. New York: Wiley-Interscience.

Wolfle, L. M. 1985. Applications of causal models in higher education. In Higher Education: Handbook of Theory and Research, vol.1, J.C. Smart, ed. New York: Agathon Press, pp. 381-413.

Zwick, R., and H. I. Brawn. 1988. Methods for Analyzing the Attainment of Graduate School Milestones: A Case Study (GRE Board Professional Report No. 86-3P/ETS Research Report No. 88-30). Princeton, N.J.: Educational Testing Service.

APPENDIXES

A: RELATED TABLES

TABLE A1: Median Time to the Doctorate

Field of Doctorate	TTD 1967	1986
Chemistry	5.27	6.08
Physics/Astronomy	6.34	7.07
Earth/Atmospheric/Marine Sciences	7.24	8.80
Math/Computer Sciences	6.03	7.89
Engineering	7.09	7.83
Agricultural Sciences	7.52	8.48
Biosciences	6.69	7.98
Health Sciences	9.64	12.10
Psychology	6.84	9.71
Economics	7.72	8.07
Social Sciences	8.61	11.50
	RTD	
Chemistry	4.70	5.39
Physics/Astronomy	5.58	6.30
Earth/Atmospheric/Marine Sciences	5.44	6.94
Math/Computer Sciences	5.11	6.29
Engineering	5.28	5.98
Agricultural Sciences	5.14	6.14
Biosciences	5.47	6.42
Health Sciences	5.60	7.06
Psychology	5.32	7.02
Economics	4.74	6.41
Social Sciences	5.32	8.27

TABLE A2.1: Mean Total Time to the Doctorate (TTD), 1967 and 1986

Field of Doctorate	1967 TTD	1967 Standard Deviation	1986 TTD	1986 Standard Deviation
Total 11 Fields	8.19	4.69	9.84	5.06
Chemistry	6.36	3.50	7.20	3.78
Physics/Astronomy	7.10	3.03	8.06	3.65
Earth/Atmospheric/Marine Sciences	8.73	4.22	9.98	4.57
Math/Computer Sciences	7.15	3.86	9.27	4.87
Engineering	8.39	4.49	9.27	4.88
Agricultural Sciences	8.75	4.31	9.49	4.17
Biosciences	8.34	4.76	8.99	4.06
Health Sciences	10.50	5.67	13.31	5.92
Psychology	8.57	5.20	10.90	5.29
Economics	9.20	5.04	9.54	5.03
Social Sciences	10.59	6.29	12.94	6.05

TABLE A2.2: Mean Time Prior to Graduate Entrance (TPGE), 1967 and 1986

Field of Doctorate	1967 TPGE	1967 Standard Deviation	1986 TPGE	1986 Standard Deviation
Total 11 Fields	0.85	2.24	0.18	2.44
Chemistry	0.51	1.70	0.67	1.98
Physics/Astronomy	0.34	1.09	0.50	1.44
Earth/Atmospheric/Marine Sciences	0.98	2.21	0.97	2.09
Math/Computer Sciences	0.44	1.37	0.90	2.01
Engineering	0.96	2.06	1.03	2.15
Agricultural Sciences	1.42	2.58	1.19	2.03
Biosciences	0.89	2.22	1.11	2.22
Health Sciences	1.04	2.32	2.08	3.24
Psychology	0.92	2.55	1.50	2.79
Economics	1.03	2.33	1.08	2.39
Social Sciences	1.42	3.67	1.54	3.11

TABLE A2.3: Mean Registered Time to the Doctorate (RTD), 1967 and 1986

	1967		1986	
Field of Doctorate	RTD	Standard Deviation	RTD	Standard Deviation
Total 11 Fields	5.63	2.03	7.02	2.52
Chemistry	5.01	1.64	5.83	1.86
Physics/Astronomy	5.90	1.76	6.69	2.09
Earth/Atmospheric/Marine Sciences	5.72	1.79	7.34	2.30
Math/Computer Sciences	5.39	1.75	6.97	2.73
Engineering	5.68	2.06	6.48	2.44
Agricultural Sciences	5.38	1.75	6.57	2.22
Biosciences	5.83	2.03	6.77	2.03
Health Sciences	6.46	2.84	7.34	2.49
Psychology	5.86	2.29	7.48	2.61
Economics	5.11	2.01	7.01	2.76
Social Sciences	5.88	2.44	8.78	3.09

TABLE A2.4: Mean Time Not Enrolled in University (TNEU), 1967 and 1986

	1967		1986	
Field of Doctorate	TNEU	Standard Deviation	TNEU	Standard Deviation
Total 11 Fields	1.67	3.08	1.59	3.12
Chemistry	0.80	1.91	0.68	1.85
Physics/Astronomy	0.87	1.93	0.77	2.10
Earth/Atmospheric/Marine Sciences	1.98	2.91	1.59	2.82
Math/Computer Sciences	1.17	2.52	1.40	2.86
Engineering	1.74	2.99	1.71	3.15
Agricultural Sciences	1.96	3.22	1.69	2.73
Biosciences	1.58	3.03	1.07	2.41
Health Sciences	2.99	4.38	3.92	4.52
Psychology	1.80	3.23	1.88	3.36
Economics	2.95	3.80	1.45	3.01
Social Sciences	3.22	4.41	2.51	4.22

TABLE A2.5: Mean Total Years Change, 1967 and 1986

Field of Doctorate	TTD	RTD	TPGE	TNEU
Chemistry	0.84	0.82	0.16	-0.12
Physics/Astronomy	0.96	0.79	0.16	-0.10
Earth/Atmospheric/Marine Sciences	1.25	1.62	-0.01	-0.39
Math/Computer Sciences	2.12	1.58	0.46	0.23
Engineering	0.88	0.80	0.07	-0.03
Agricultural Sciences	0.74	1.19	-0.23	-0.27
Biosciences	0.65	0.94	0.22	-0.51
Health Sciences	2.81	0.88	1.04	0.93
Psychology	2.33	1.62	0.58	0.08
Economics	0.34	1.90	0.05	-1.50
Social Sciences	2.35	2.90	0.12	-0.71

Note: The sum of the changes in these fields may not equal the change in total time to the doctorate because of rounding errors and missing data in some fields.

TABLE A3: Estimated Personnel Loss, by Field, 1967 to 1986

Field Table 3.1: Chemistry

Year of Graduation	Total Doctorates	Mean Total TTD	Person Year Loss	Loss as % of Total	Change of TTD
1967	1548	6.36	-	-	-
1968	1594	6.40	64	4	0.04
1969	1753	6.42	105	6	0.06
1970	2038	6.55	387	19	0.19
1971	2011	6.49	261	13	0.13
1972	1808	6.87	922	51	0.51
1973	1633	6.91	898	55	0.55
1974	1542	6.85	756	49	0.49
1975	1519	6.85	744	49	0.49
1976	1405	6.98	871	62	0.62
1977	1343	6.91	739	55	0.55
1978	1293	6.96	776	60	0.6
1979	1335	6.71	467	35	0.35
1980	1269	6.68	406	32	0.32
1981	1329	6.68	425	32	0.32
1982	1369	6.58	301	22	0.22
1983	1424	6.97	869	61	0.61
1984	1415	6.94	821	58	0.58
1985	1432	6.93	816	57	0.57
1986	1412	7.20	1186	84	0.84
TOTAL	28924		11815	41	

Field Table 3.2: Physics/Astronomy

Year of Graduation	Total Doctorates	Mean Total TTD	Person Year Loss	Loss as % of Total	Change of TTD
1967	1119	7.10	-	-	-
1968	1249	7.18	100	8	0.08
1969	1258	7.18	101	8	0.08
1970	1436	7.21	158	11	0.11
1971	1503	7.49	586	39	0.39
1972	1403	7.49	547	39	0.39
1973	1314	7.83	959	73	0.73
1974	1054	7.99	938	89	0.89
1975	1034	7.97	900	87	0.87
1976	987	7.97	859	87	0.87
1977	919	8.12	937	102	1.02
1978	868	8.02	799	92	0.92
1979	870	7.75	566	65	0.65
1980	766	7.92	628	82	0.82
1981	774	7.73	488	63	0.63
1982	741	8.03	689	93	0.93
1983	760	7.86	578	76	0.76
1984	775	7.82	558	72	0.72
1985	746	8.05	709	95	0.95
1986	732	8.06	703	96	0.96
TOTAL	19189		11801	61	

Field Table 3.3: Earth, Atmospheric, and Marine Sciences

Year of Graduation	Total Doctorates	Mean Total TTD	Person Year Loss	Loss as % of Total	Change of TTD
1967	347	8.73	-	-	-
1968	367	8.71	-7	-2	-0.02
1969	418	9.09	150	36	0.36
1970	433	8.85	52	12	0.12
1971	506	9.14	207	41	0.41
1972	531	9.27	287	54	0.54
1973	554	9.11	211	38	0.38
1974	504	9.32	297	59	0.59
1975	530	9.14	217	41	0.41
1976	540	8.48	-135	-25	-0.25
1977	581	9.02	168	29	0.29
1978	540	8.67	-32	-6	-0.06
1979	566	8.60	-74	-13	-0.13
1980	538	9.24	274	51	0.51
1981	488	9.22	239	49	0.49
1982	557	9.26	295	53	0.53
1983	513	9.48	385	75	0.75
1984	499	9.56	414	83	0.83
1985	474	9.50	365	77	0.77
1986	446	9.98	558	125	1.25
TOTAL	9585		3872	40	

Field Table 3.4: Math/Computer Science

Year of Graduation	Total Doctorates	Mean Total TTD	Person Year Loss	Loss as % of Total	Change of TTD
1967	719	7.15	-	-	-
1968	841	7.01	-118	-14	-0.14
1969	937	7.05	-94	-10	-0.1
1970	1076	7.08	-75	-7	-0.07
1971	1074	7.40	269	25	0.25
1972	1095	7.85	767	70	0.7
1973	1033	7.67	537	52	0.52
1974	947	8.11	909	96	0.96
1975	923	7.79	591	64	0.64
1976	803	7.99	675	84	0.84
1977	769	7.92	592	77	0.77
1978	756	8.24	824	109	1.09
1979	778	8.36	941	121	1.21
1980	751	8.17	766	102	1.02
1981	713	8.26	791	111	1.11
1982	654	8.20	687	105	1.05
1983	664	8.96	1202	181	1.81
1984	638	9.26	1346	211	2.11
1985	631	9.23	1312	208	2.08
1986	653	9.27	1384	212	2.12
TOTAL	15736		13306	85	

Field Table 3.5: Engineering

Year of Graduation	Total Doctorates	Mean Total TTD	Person Year Loss	Loss as % of Total	Change of TTD
1967	2155	8.39	-	-	-
1968	2378	8.23	-380	-16	-0.16
1969	2736	8.26	-356	-13	-0.13
1970	2944	8.16	-677	-23	-0.23
1971	2948	8.45	177	6	0.06
1972	2952	8.78	1151	39	0.39
1973	2699	9.18	2132	79	0.79
1974	2267	8.97	1315	58	0.58
1975	2134	8.90	1088	51	0.51
1976	1947	9.23	1635	84	0.84
1977	1798	8.93	971	54	0.54
1978	1586	8.90	809	51	0.51
1979	1615	8.84	727	45	0.45
1980	1554	8.99	932	60	0.6
1981	1471	9.30	1339	91	0.91
1982	1465	9.21	1201	82	0.82
1983	1482	8.99	889	60	0.6
1984	1513	8.98	893	59	0.59
1985	1595	9.05	1053	66	0.66
1986	1722	9.27	1515	88	0.8
TOTAL	38806		16415	42	

Field Table 3.6: Agricultural Sciences

Year of Graduation	Total Doctorates	Mean Total TTD	Person Year Loss	Loss as % of Total	Change of TTD
1967	421	8.75	-	-	-
1968	479	9.16	196	41	0.41
1969	582	8.52	-134	-23	-0.23
1970	683	8.51	-164	-24	-0.24
1971	782	8.83	63	8	0.08
1972	729	8.99	175	24	0.24
1973	737	8.79	29	4	0.04
1974	644	8.75	0	0	0.00
1975	736	8.92	125	17	0.17
1976	636	9.22	299	47	0.47
1977	601	8.79	24	4	0.04
1978	666	8.52	-153	-23	-0.23
1979	652	8.45	-196	-30	-0.3
1980	708	8.49	-184	-26	-0.26
1981	732	8.39	-264	-36	-0.36
1982	773	8.41	-263	-34	-0.34
1983	773	8.72	-23	-3	-0.03
1984	742	8.93	134	18	0.18
1985	813	9.13	309	38	0.38
1986	711	9.49	526	74	0.74
TOTAL	13179		500	4	

Field Table 3.7: Biosciences

Year of Graduation	Total Doctorates	Mean Total TTD	Person Year Loss	Loss as % of Total	Change of TTD
1967	2026	8.34	-	-	-
1968	2436	8.08	-633	-26	-0.26
1969	2712	7.78	-1519	-56	-0.56
1970	2975	7.52	-2440	-82	-0.82
1971	3264	7.61	-2383	-73	-0.73
1972	3216	7.86	-1544	-48	-0.48
1973	3258	7.93	-1336	-41	-0.41
1974	2957	7.88	-1360	-46	-0.46
1975	3100	7.84	-1550	-50	-0.50
1976	3160	7.99	-1106	-35	-0.35
1977	3071	7.92	-1290	-42	-0.42
1978	3134	7.86	-1504	-48	-0.48
1979	3262	7.95	-1272	-39	-0.39
1980	3430	7.91	-1475	-43	-0.43
1981	3421	7.93	-1403	-41	-0.41
1982	3434	8.09	-859	-25	-0.25
1983	3323	8.38	133	4	0.04
1984	3399	8.67	1122	33	0.33
1985	3246	8.72	1233	38	0.38
1986	3234	8.99	2102	65	0.65
TOTAL	60032		-17082	-28	

Field Table 3.8: Health Science

Year of Graduation	Total Doctorates	Mean Total TTD	Person Year Loss	Loss as % of Total	Change of TTD
1967	140	10.50	-	-	-
1968	165	10.96	76	46	0.46
1969	240	10.47	-7	-3	-0.03
1970	350	10.52	7	2	0.02
1971	479	10.17	-158	-33	-0.33
1972	407	9.72	-317	-78	-0.78
1973	432	9.77	-315	-73	-0.73
1974	407	10.19	-126	-31	-0.31
1975	397	10.65	60	15	0.15
1976	434	10.50	0	0	0.00
1977	461	10.67	78	17	0.17
1978	444	10.29	-93	-21	-0.21
1979	491	10.47	-15	-3	-0.03
1980	506	10.76	132	26	0.26
1981	586	11.25	440	75	0.75
1982	587	11.82	775	132	1.32
1983	531	11.91	749	141	1.41
1984	617	12.16	1024	166	1.66
1985	585	13.12	1533	262	2.62
1986	601	13.31	1689	281	2.81
TOTAL	8720		5529	63	

Field Table 3.9: Psychology

Year of Graduation	Total Doctorates	Mean Total TTD	Person Year Loss	Loss as % of Total	Change of TTD
1967	1240	8.57	-	-	-
1968	1406	8.62	70	5	0.05
1969	1671	8.38	-317	-19	-0.19
1970	1816	8.09	-872	-48	-0.48
1971	2042	7.96	-1246	-61	-0.61
1972	2169	8.24	-716	-33	-0.33
1973	2335	8.38	-444	-19	-0.19
1974	2391	8.44	-311	-13	-0.13
1975	2607	8.42	-391	-15	-0.15
1976	2768	8.35	-609	-22	-0.22
1977	2821	8.63	169	6	0.06
1978	2858	8.72	429	15	0.15
1979	2895	9.05	1390	48	0.48
1980	2909	9.27	2036	70	0.7
1981	3158	9.63	3347	106	1.06
1982	2922	9.98	4120	141	1.41
1983	3090	10.17	4944	160	1.6
1984	2965	10.49	5693	192	1.92
1985	2837	10.73	6128	216	2.16
1986	2796	10.90	6515	233	2.33
TOTAL	48456		29936	62	

Field Table 3.10: Economics

Year of Graduation	Total Doctorates	Mean Total TTD	Person Year Loss	Loss as % of Total	Change of TTD
1967	546	9.20	-	-	-
1968	582	9.50	175	30	0.3
1969	572	8.96	-137	-24	-0.24
1970	695	8.86	-236	-34	-0.34
1971	661	8.64	-370	-56	-0.56
1972	711	8.75	-320	-45	-0.45
1973	729	8.80	-292	-40	-0.4
1974	633	8.46	-468	-74	-0.74
1975	692	8.53	-464	-67	-0.67
1976	668	8.90	-200	-30	-0.3
1977	610	8.65	-336	-55	-0.55
1978	584	9.09	-64	-11	-0.1
1979	580	9.15	-29	-5	-0.05
1980	557	9.25	28	5	0.05
1981	588	9.08	-71	-12	-0.12
1982	507	9.38	91	18	0.18
1983	564	9.56	203	36	0.36
1984	512	9.57	189	37	0.37
1985	499	9.67	235	47	0.47
1986	533	9.54	181	34	0.34
TOTAL	11477		-1885	-16	

Field Table 3.11: Social Sciences

Year of Graduation	Total Doctorates	Mean Total TTD	Person Year Loss	Loss as % of Total	Change of TTD
1967	957	10.59	-	-	-
1968	1122	10.69	112	10	0.1
1969	1316	10.29	-395	-30	-0.3
1970	1599	10.34	-400	-25	-0.25
1971	1993	10.04	-1096	-55	-0.55
1972	2012	9.94	-1308	-65	-0.65
1973	2058	10.37	-453	-22	-0.22
1974	2053	10.27	-657	-32	-0.32
1975	2097	10.32	-566	-27	-0.27
1976	2124	10.18	-871	-41	-0.41
1977	1940	10.54	-97	-5	-0.05
1978	1888	10.48	-208	-11	-0.11
1979	1780	11.02	765	43	0.43
1980	1722	10.86	465	27	0.27
1981	1621	11.18	956	59	0.59
1982	1567	11.65	1661	106	1.06
1983	1566	11.88	2020	129	1.29
1984	1479	12.29	2514	170	1.7
1985	1427	12.64	2925	205	2.05
1986	1439	12.94	3382	235	2.35
TOTAL	32803		8751	27	

TABLE A4: Descriptive Statistics of Variables in the Data Base, by Field, 1967 to 1986

Field Table 4.1: Chemistry

Variable	Mean of Years	Standard Deviation	Minimum	Maximum
Number of Ph.D.s	1523.6	218.9	1269.0	2038.0
Ph.D.s per 100,000 of U.S. Population	26.8	14.1	14.0	51.9
Total Time to the Doctorate	6.8	0.2	6.4	7.2
Registered Time to the Doctorate	5.5	0.2	5.0	5.8
Percent with Federal Support	34.2	15.3	11.6	55.0
Percent with Private Support	2.6	1.0	1.4	5.2
Percent with T.A.	80.0	3.5	72.7	85.2
Percent with R.A.	67.6	8.5	55.8	81.6
Percent with Spousal Support	21.5	8.6	10.5	38.3
Percent Married	66.0	9.2	52.1	76.9
Mean Number of Dependents	1.0	0.3	0.7	1.6
Percent with Foreign Baccalaureate	6.9	1.9	3.2	10.2
Median Age Starting Ph.D.	22.3	0.1	22.1	22.4
Percent with Definite Employment/Study*	83.9	6.1	74.4	93.6
Percent Negotiating Employment/Study*	5.5	1.3	3.4	8.7
Percent seeking Employment/Study*	10.4	5.1	2.4	19.5
Percent with Temporary Visas	13.2	3.7	7.9	21.8
Percent with Postdoc Study Plan*	40.4	8.4	28.1	54.0
Percent with Same B.S. Fields	88.2	1.3	85.7	90.3
Percent of Women Ph.D.s	12.5	4.6	6.3	21.9

Male Ph.D.s Time to the Doctorate	5.8	0.3	5.3	6.2
Female Ph.D.s Time to the Doctorate	6.0	0.2	5.5	6.4
Percent with B.A. from Top 20 NRC	11.6	1.6	9.0	15.6
Percent with B.A. from Top 40 NRC	19.5	1.7	17.2	24.6
Percent of Ph.D.s at Top 20 NRC	32.7	2.4	28.7	39.3
Percent of Ph.D.s at Top 40 NRC	50.9	2.6	45.8	56.5
Black Ph.D.s Time to the Doctorate	7.7	1.0	5.8	9.5
White Ph.D.s Time to the Doctorate	5.8	0.2	5.5	6.0
Percent with Primary Support Personal**	4.3	4.5	0.0	11.3
Percent with Primary Support Own Earning**	2.5	1.6	0.0	5.1
Percent of Black Ph.D.s***	1.6	0.4	1.0	2.4
Percent of Hispanic Ph.D.s***	1.4	0.5	0.7	2.4
Percent with Bacc from Research I	29.9	1.8	27.1	33.6
Percent with Bacc from Research I & II	38.8	1.7	36.3	42.7
Percent of Ph.D.s at Research I	66.3	2.3	62.5	69.9
Percent of Ph.D.s at Research I & II	83.0	1.5	80.6	85.4

NOTES: Top 20 and Top 40 NRC refer to the National Research Council's 1981-82 Assessment of Research-Doctorate Programs in the United States. Research I and Research II are obtained from the Carnegie Classification of Colleges and Universities.

*Data from 1964-1983 **Data from 1977-1986 ***Data from 1974-1986

Field Table 4.2: Physics/Astronomy

Variable	Mean of Years	Standard Deviation	Minimum	Maximum
Number of Ph.D.s	1015.4	255.4	732.0	1503.0
Ph.D.s per 100,000 of U.S. Population	18.6	11.4	7.3	36.6
Total Time to the Doctorate	7.7	0.3	7.1	8.1
Registered Time to the Doctorate	6.5	0.3	5.9	6.9
Percent with Federal Support	35.3	15.1	11.3	52.7
Percent with Private Support	3.2	1.1	1.5	5.7
Percent with T.A.	65.2	3.8	58.5	70.4
Percent with R.A.	77.0	4.8	70.7	85.1
Percent with Spousal Support	19.8	6.9	10.1	31.6
Percent Married	64.2	11.0	48.1	78.2
Mean Number of Dependents	1.1	0.4	0.6	1.7
Percent with Foreign Baccalaureate	6.6	1.7	3.9	9.4
Median Age Starting Ph.D.	22.2	0.1	22.1	22.3
Percent with Definite Employment/Study*	76.0	5.2	66.2	83.3
Percent Negotiating Employment/Study*	10.0	1.2	6.9	11.8
Percent seeking Employment/Study*	13.7	5.7	4.3	23.9
Percent with Temporary Visas	18.7	6.1	11.0	33.3
Percent with Postdoc Study Plan*	39.5	11.1	18.7	50.4
Percent with Same B.S. Fields	81.1	3.2	72.5	84.7
Percent of Women Ph.D.s	5.1	2.3	2.3	9.1

Male Ph.D.s Time to the Doctorate	6.9	0.3	6.3	7.3
Female Ph.D.s Time to the Doctorate	7.1	0.4	6.2	7.7
Percent with B.A. from Top 20 NRC	23.0	1.9	19.5	26.3
Percent with B.A. from Top 40 NRC	33.0	1.9	28.9	35.7
Percent of Ph.D.s at Top 20 NRC	38.1	4.0	32.0	44.3
Percent of Ph.D.s at Top 40 NRC	57.8	3.0	53.0	62.2
Black Ph.D.s Time to the Doctorate	8.5	1.1	7.0	10.0
White Ph.D.s Time to the Doctorate	7.0	0.2	6.7	7.2
Percent with Primary Support Personal**	4.6	4.8	0.0	11.7
Percent with Primary Support Own Earning**	2.9	2.5	0.0	6.7
Percent of Black Ph.D.s***	0.9	0.4	0.3	1.5
Percent of Hispanic Ph.D.s***	1.3	0.5	0.7	2.1
Percent with Bacc from Research I	47.1	2.4	42.5	50.8
Percent with Bacc from Research I & II	56.0	2.5	50.3	60.7
Percent of Ph.D.s at Research I	74.4	2.9	69.6	79.5
Percent of Ph.D.s at Research I & II	86.3	1.9	82.9	89.6

NOTES: Top 20 and Top 40 NRC refer to the National Research Council's 1981-82 Assessment of Research-Doctorate Programs in the United States. Research I and Research II are obtained from the Carnegie Classification of Colleges and Universities.

*Data from 1964-1983 **Data from 1977-1986 ***Data from 1974-1986

Field Table 4.3: Earth, Atmospheric, and Marine Sciences

Variable	Mean of Years	Standard Deviation	Minimum	Maximum
Number of Ph.D.s	496.6	63.1	347.0	581.0
Ph.D.s per 100,000 of U.S. Population	8.1	2.5	4.4	12.1
Total Time to the Doctorate	9.1	0.4	8.5	10.0
Registered Time to the Doctorate	6.5	0.5	5.7	7.3
Percent with Federal Support	42.5	13.5	18.8	62.2
Percent with Private Support	3.1	1.1	1.3	6.0
Percent with T.A.	52.9	3.1	47.2	61.7
Percent with R.A.	62.2	9.7	46.0	74.1
Percent with Spousal Support	27.4	7.8	15.8	42.7
Percent Married	72.7	10.4	58.8	87.0
Mean Number of Dependents	1.4	0.5	0.8	2.2
Percent with Foreign Baccalaureate	6.1	0.8	3.5	10.2
Median Age Starting Ph.D.	22.6	0.2	22.4	22.9
Percent with Definite Employment/Study*	81.2	4.2	74.7	91.2
Percent Negotiating Employment/Study*	9.6	1.2	7.3	11.4
Percent seeking Employment/Study*	9.0	3.7	1.5	16.4
Percent with Temporary Visas	14.4	2.9	8.2	20.1
Percent with Postdoc Study Plan*	21.6	7.9	7.1	32.5
Percent with Same B.S. Fields	53.9	6.2	48.0	69.2
Percent of Women Ph.D.s	9.3	5.9	1.4	19.4

Male Ph.D.s Time to the Doctorate	8.0	0.4	7.2	8.7
Female Ph.D.s Time to the Doctorate	7.9	0.8	6.0	9.2
Percent with B.A. from Top 20 NRC	16.0	1.6	13.5	19.5
Percent with B.A. from Top 40 NRC	28.5	2.5	24.3	32.9
Percent of Ph.D.s at Top 20 NRC	30.2	2.4	24.9	34.6
Percent of Ph.D.s at Top 40 NRC	56.2	2.8	49.3	60.3
Black Ph.D.s Time to the Doctorate				
White Ph.D.s Time to the Doctorate	8.0	0.4	7.4	8.7
Percent with Primary Support Personal**	8.6	8.8	0.0	21.8
Percent with Primary Support Own Earning**	6.9	5.6	0.0	15.5
Percent of Black Ph.D.s***	0.4	0.3	0.0	0.9
Percent of Hispanic Ph.D.s***	1.1	0.6	0.0	2.2
Percent with Bacc from Research I	41.8	2.8	38.2	48.6
Percent with Bacc from Research I & II	52.7	2.7	48.3	58.6
Percent of Ph.D.s at Research I	74.5	3.7	69.3	83.6
Percent of Ph.D.s at Research I & II	88.3	1.9	85.5	93.7

NOTES: Top 20 and Top 40 NRC refer to the National Research Council's 1981-82 Assessment of Research-Doctorate Programs in the United States. Research I and Research II are obtained from the Carnegie Classification of Colleges and Universities.

*Data from 1964-1983 **Data from 1977-1986 ***Data from 1974-1986

Field Table 4.4: Math/Computer Sciences

Variable	Mean of Years	Standard Deviation	Minimum	Maximum
Number of Ph.D.s	822.8	154.2	631.0	1095.0
Ph.D.s per 100,000 of U.S. Population	14.6	7.6	6.5	27.4
Total Time to the Doctorate	8.0	0.7	7.0	9.3
Registered Time to the Doctorate	6.2	0.5	5.4	7.0
Percent with Federal Support	38.3	18.8	13.9	62.3
Percent with Private Support	3.5	1.5	1.1	6.4
Percent with T.A.	73.5	3.9	66.5	79.0
Percent with R.A.	41.1	5.8	31.3	50.8
Percent with Spousal Support	21.8	6.6	13.1	35.2
Percent Married	63.3	9.1	51.2	76.3
Mean Number of Dependents	1.1	0.3	0.7	1.5
Percent with Foreign Baccalaureate	7.3	2.2	4.3	12.6
Median Age Starting Ph.D.	22.2	0.1	22.1	22.4
Percent with Definite Employment/Study*	77.2	5.3	66.1	84.5
Percent Negotiating Employment/Study*	9.2	1.1	7.2	11.1
Percent seeking Employment/Study*	13.1	6.0	4.3	23.6
Percent with Temporary Visas	20.1	8.1	10.8	37.6
Percent with Postdoc Study Plan*	8.4	2.5	4.1	14.5
Percent with Same B.S. Fields	71.9	7.1	56.8	80.6
Percent of Women Ph.D.s	11.8	4.4	5.1	17.9

Male Ph.D.s Time to the Doctorate	6.9	0.6	5.9	8.0
Female Ph.D.s Time to the Doctorate	7.4	0.7	6.4	8.6
Percent with B.A. from Top 20 NRC	21.3	1.9	17.5	24.3
Percent with B.A. from Top 40 NRC	30.8	2.5	26.7	35.2
Percent of Ph.D.s at Top 20 NRC	38.5	2.6	32.6	44.2
Percent of Ph.D.s at Top 40 NRC	57.2	3.5	50.8	64.7
Black Ph.D.s Time to the Doctorate	10.9	2.1	8.3	15.3
White Ph.D.s Time to the Doctorate	7.1	0.4	6.7	8.0
Percent with Primary Support Personal**	9.5	9.8	0.0	24.7
Percent with Primary Support Own Earning**	12.2	2.8	9.4	18.6
Percent of Black Ph.D.s***	1.4	0.4	0.7	2.1
Percent of Hispanic Ph.D.s***	1.4	0.8	0.3	3.0
Percent with Bacc from Research I	42.1	2.9	37.9	49.4
Percent with Bacc from Research I & II	51.4	2.9	47.1	57.2
Percent of Ph.D.s at Research I	71.3	1.9	67.6	74.8
Percent of Ph.D.s at Research I & II	85.5	2.2	81.7	91.4

NOTES: Top 20 and Top 40 NRC refer to the National Research Council's 1981-82 Assessment of Research-Doctorate Programs in the United States. Research I and Research II are obtained from the Carnegie Classification of Colleges and Universities.

*Data from 1964-1983 **Data from 1977-1986 ***Data from 1974-1986

Field Table 4.5: Engineering

Variable	Mean of Years	Standard Deviation	Minimum	Maximum
Number of Ph.D.s	2048.1	539.0	1465.0	2952.0
Ph.D.s per 100,000 of U.S. Population	37.6	23.2	15.4	75.0
Total Time to the Doctorate	8.9	0.4	8.2	9.3
Registered Time to the Doctorate	6.2	0.3	5.5	6.6
Percent with Federal Support	37.1	13.3	14.5	54.5
Percent with Private Support	2.8	1.5	1.5	6.5
Percent with T.A.	39.4	2.3	35.4	44.8
Percent with R.A.	60.8	7.2	49.6	70.7
Percent with Spousal Support	21.1	7.1	11.6	34.1
Percent Married	75.8	7.4	63.9	86.1
Mean Number of Dependents	1.6	0.4	1.1	2.2
Percent with Foreign Baccalaureate	17.6	3.3	9.9	27.8
Median Age Starting Ph.D.	22.8	0.1	22.6	23.0
Percent with Definite Employment/Study*	80.2	4.6	72.4	89.6
Percent Negotiating Employment/Study*	10.4	1.0	8.0	12.2
Percent seeking Employment/Study*	9.1	4.3	2.3	16.8
Percent with Temporary Visas	29.1	11.7	13.8	47.1
Percent with Postdoc Study Plan*	8.7	2.9	3.7	12.4
Percent with Same B.S. Fields	81.1	6.2	73.2	90.5
Percent of Women Ph.D.s	3.1	2.8	0.3	9.2

Male Ph.D.s Time to the Doctorate	7.7	0.4	6.9	8.2
Female Ph.D.s Time to the Doctorate	7.5	0.8	6.2	9.5
Percent with B.A. from Top 20 NRC	22.9	2.3	18.0	27.3
Percent with B.A. from Top 40 NRC	36.1	2.9	32.1	42.8
Percent of Ph.D.s at Top 20 NRC	43.6	2.7	39.2	47.7
Percent of Ph.D.s at Top 40 NRC	64.2	2.6	60.4	70.4
Black Ph.D.s Time to the Doctorate	8.7	1.0	7.4	11.0
White Ph.D.s Time to the Doctorate	7.6	0.2	7.3	7.9
Percent with Primary Support Personal**	11.0	10.0	0.0	22.9
Percent with Primary Support Own Earning**	9.5	7.4	0.0	17.7
Percent of Black Ph.D.s***	1.3	0.4	0.8	2.2
Percent of Hispanic Ph.D.s***	1.6	0.6	0.7	2.6
Percent with Bacc from Research I	52.5	1.8	50.2	57.1
Percent with Bacc from Research I & II	64.6	1.2	62.8	67.1
Percent of Ph.D.s at Research I	74.7	1.6	72.4	77.3
Percent of Ph.D.s at Research I & II	86.8	1.3	85.0	89.6

NOTES: Top 20 and Top 40 NRC refer to the National Research Council's 1981-82 Assessment of Research-Doctorate Programs in the United States. Research I and Research II are obtained from the Carnegie Classification of Colleges and Universities.

*Data from 1964-1983 **Data from 1977-1986 ***Data from 1974-1986

Field Table 4.6: Agricultural Sciences

Variable	Mean of Years	Standard Deviation	Minimum	Maximum
Number of Ph.D.s	680.0	97.5	421.0	813.0
Ph.D.s per 100,000 of U.S. Population	11.0	3.7	7.0	18.8
Total Time to the Doctorate	8.8	0.3	8.4	9.5
Registered Time to the Doctorate	5.9	0.3	5.2	6.6
Percent with Federal Support	27.9	11.7	6.8	43.5
Percent with Private Support	1.8	0.9	0.6	4.4
Percent with T.A.	24.7	2.3	20.3	28.9
Percent with R.A.	77.6	3.6	70.3	81.9
Percent with Spousal Support	32.6	10.3	17.8	47.4
Percent Married	80.8	7.9	66.7	90.8
Mean Number of Dependents	1.8	0.5	1.1	2.5
Percent with Foreign Baccalaureate	7.5	2.4	4.0	12.4
Median Age Starting Ph.D.	23.2	0.2	22.9	24.0
Percent with Definite Employment/Study*	77.4	6.8	68.7	92.3
Percent Negotiating Employment/Study*	9.9	1.7	6.0	12.5
Percent seeking Employment/Study*	12.4	5.5	1.4	18.7
Percent with Temporary Visas	31.3	2.8	25.4	34.8
Percent with Postdoc Study Plan*	11.9	3.6	4.9	17.5
Percent with Same B.S. Fields	62.3	9.6	46.6	77.9
Percent of Women Ph.D.s	8.2	6.2	1.0	19.0

Male Ph.D.s Time to the Doctorate	7.5	0.4	6.7	8.5
Female Ph.D.s Time to the Doctorate	7.8	1.2	4.2	10.5
Percent with B.A. from Top 20 NRC	12.0	1.9	8.6	15.9
Percent with B.A. from Top 40 NRC	18.3	2.5	14.1	24.0
Percent of Ph.D.s at Top 20 NRC	23.2	2.2	18.7	27.0
Percent of Ph.D.s at Top 40 NRC	35.7	3.8	26.9	44.3
Black Ph.D.s Time to the Doctorate	8.5	1.3	6.3	11.0
White Ph.D.s Time to the Doctorate	7.5	0.3	7.1	8.4
Percent with Primary Support Personal**	10.0	10.4	0.0	25.6
Percent with Primary Support Own Earning**	8.8	5.1	0.0	13.8
Percent of Black Ph.D.s***	2.0	0.7	0.7	2.9
Percent of Hispanic Ph.D.s***	1.4	0.7	0.2	2.7
Percent with Bacc from Research I	46.2	2.8	42.1	52.9
Percent with Bacc from Research I & II	65.0	4.2	58.1	74.1
Percent of Ph.D.s at Research I	75.0	2.1	69.5	79.3
Percent of Ph.D.s at Research I & II	93.4	1.2	90.0	95.0

NOTES: Top 20 and Top 40 NRC refer to the National Research Council's 1981-82 Assessment of Research-Doctorate Programs in the United States. Research I and Research II are obtained from the Carnegie Classification of Colleges and Universities.

*Data from 1964-1983 **Data from 1977-1986 ***Data from 1974-1986

Field Table 4.7: Biosciences

Variable	Mean of Years	Standard Deviation	Minimum	Maximum
Number of Ph.D.s	3102.9	345.0	2026.0	3434.0
Ph.D.s per 100,000 of U.S. Population	50.5	15.8	32.0	78.3
Total Time to the Doctorate	8.1	0.4	7.5	9.0
Registered Time to the Doctorate	6.1	0.3	5.7	6.8
Percent with Federal Support	57.3	12.4	38.4	73.3
Percent with Private Support	2.3	1.0	1.3	5.1
Percent with T.A.	47.6	3.9	37.9	52.4
Percent with R.A.	42.4	6.5	34.7	54.0
Percent with Spousal Support	25.8	8.6	13.8	39.0
Percent Married	67.8	9.3	56.0	80.4
Mean Number of Dependents	1.1	0.4	0.7	1.9
Percent with Foreign Baccalaureate	4.5	1.2	2.8	7.0
Median Age Starting Ph.D.	22.5	0.1	22.4	22.8
Percent with Definite Employment/Study*	80.3	4.2	75.2	87.7
Percent Negotiating Employment/Study*	8.5	1.0	6.6	10.3
Percent seeking Employment/Study*	10.7	3.5	4.7	14.5
Percent with Temporary Visas	9.7	1.5	7.8	12.9
Percent with Postdoc Study Plan*	53.4	13.2	32.6	70.6
Percent with Same B.S. Fields	57.9	3.6	49.9	63.8
Percent of Women Ph.D.s	24.4	6.1	15.5	34.4

Male Ph.D.s Time to the Doctorate	6.9	0.5	6.3	7.9
Female Ph.D.s Time to the Doctorate	7.2	0.5	6.3	8.2
Percent with B.A. from Top 20 NRC	14.8	1.8	11.6	17.9
Percent with B.A. from Top 40 NRC	23.7	2.5	19.5	27.9
Percent of Ph.D.s at Top 20 NRC	26.8	1.5	25.0	30.8
Percent of Ph.D.s at Top 40 NRC	42.7	2.4	39.5	48.5
Black Ph.D.s Time to the Doctorate	9.0	1.0	7.8	10.9
White Ph.D.s Time to the Doctorate	7.1	0.4	6.6	8.0
Percent with Primary Support Personal**	7.8	8.0	0.0	20.0
Percent with Primary Support Own Earning**	5.3	3.1	0.0	8.9
Percent of Black Ph.D.s***	1.6	0.2	1.1	1.9
Percent of Hispanic Ph.D.s***	1.2	0.4	0.8	2.1
Percent with Bacc from Research I	37.7	3.0	32.5	42.1
Percent with Bacc from Research I & II	48.2	3.1	42.6	52.5
Percent of Ph.D.s at Research I	67.1	2.2	64.9	72.9
Percent of Ph.D.s at Research I & II	83.3	3.0	79.8	90.1

NOTES: Top 20 and Top 40 NRC refer to the National Research Council's 1981-82 Assessment of Research-Doctorate Programs in the United States. Research I and Research II are obtained from the Carnegie Classification of Colleges and Universities.

*Data from 1964-1983 **Data from 1977-1986 ***Data from 1974-1986

Field Table 4.8: Health Sciences

Variable	Mean of Years	Standard Deviation	Minimum	Maximum
Number of Ph.D.s	443.0	133.4	140.0	617.0
Ph.D.s per 100,000 of U.S. Population	6.7	1.6	4.4	11.5
Total Time to the Doctorate	11.0	1.0	9.7	13.3
Registered Time to the Doctorate	6.5	0.4	5.9	7.3
Percent with Federal Support	59.2	11.5	42.7	77.2
Percent with Private Support	4.9	3.3	2.1	15.4
Percent with T.A.	28.3	3.5	18.7	34.0
Percent with R.A.	24.9	5.0	16.9	34.6
Percent with Spousal Support	27.7	11.0	8.4	41.2
Percent Married	71.5	6.9	61.7	84.2
Mean Number of Dependents	1.5	0.5	1.0	2.5
Percent with Foreign Baccalaureate	7.3	1.6	5.0	11.9
Median Age Starting Ph.D.	23.4	0.4	23.0	24.3
Percent with Definite Employment/Study*	81.0	4.2	73.5	90.8
Percent Negotiating Employment/Study*	10.6	2.4	5.8	13.7
Percent seeking Employment/Study*	7.8	3.4	0.9	13.4
Percent with Temporary Visas	11.6	3.0	7.1	18.1
Percent with Postdoc Study Plan*	13.6	4.6	4.6	19.8
Percent with Same B.S. Fields	40.5	6.7	28.5	54.7
Percent of Women Ph.D.s	36.9	17.6	7.9	70.0

Male Ph.D.s Time to the Doctorate	8.8	1.0	7.7	10.8
Female Ph.D.s Time to the Doctorate	11.6	1.3	10.1	15.0
Percent with B.A. from Top 20 NRC	13.0	2.9	9.2	21.6
Percent with B.A. from Top 40 NRC	21.0	3.1	16.7	30.3
Percent of Ph.D.s at Top 20 NRC	30.0	6.7	23.2	48.6
Percent of Ph.D.s at Top 40 NRC	51.9	6.2	43.3	71.5
Black Ph.D.s Time to the Doctorate	10.9	2.2	6.4	15.0
White Ph.D.s Time to the Doctorate	9.8	1.4	8.1	12.4
Percent with Primary Support Personal**	20.0	20.6	0.0	52.6
Percent with Primary Support Own Earning**	22.6	11.4	0.0	35.9
Percent of Black Ph.D.s***	3.7	0.9	2.1	5.4
Percent of Hispanic Ph.D.s***	1.4	0.7	0.2	2.4
Percent with Bacc from Research I	37.6	4.1	32.5	52.2
Percent with Bacc from Research I & II	48.2	3.8	43.3	59.7
Percent of Ph.D.s at Research I	74.0	4.8	66.6	91.5
Percent of Ph.D.s at Research I & II	86.4	5.1	78.3	97.6

NOTES: Top 20 and Top 40 NRC refer to the National Research Council's 1981-82 Assessment of Research-Doctorate Programs in the United States. Research I and Research II are obtained from the Carnegie Classification of Colleges and Universities.

*Data from 1964-1983 **Data from 1977-1986 ***Data from 1974-1986

Field Table 4.9: Psychology

Variable	Mean of Years	Standard Deviation	Minimum	Maximum
Number of Ph.D.s	2484.8	561.6	1240.0	3158.0
Ph.D.s per 100,000 of U.S. Population	38.2	6.2	27.7	49.0
Total Time to the Doctorate	9.1	1.0	8.0	11.0
Registered Time to the Doctorate	6.4	0.6	5.8	7.5
Percent with Federal Support	47.9	14.5	23.1	65.6
Percent with Private Support	3.2	1.8	1.4	8.0
Percent with T.A.	48.5	2.3	44.3	53.7
Percent with R.A.	38.1	3.7	32.4	44.0
Percent with Spousal Support	30.7	6.4	22.0	41.2
Percent Married	65.3	8.6	54.4	77.5
Mean Number of Dependents	1.0	0.3	0.7	1.6
Percent with Foreign Baccalaureate	1.8	0.3	1.3	2.4
Median Age Starting Ph.D.	22.7	0.3	22.4	23.3
Percent with Definite Employment/Study*	76.4	6.0	68.4	86.2
Percent Negotiating Employment/Study*	10.2	0.9	8.3	11.8
Percent seeking Employment/Study*	12.8	5.4	4.2	19.9
Percent with Temporary Visas	2.9	0.5	2.1	3.7
Percent with Postdoc Study Plan*	14.9	2.2	10.7	18.4
Percent with Same B.S. Fields	68.8	0.9	67.1	69.8
Percent of Women Ph.D.s	35.6	10.3	20.3	51.8

Male Ph.D.s Time to the Doctorate	7.4	1.1	6.1	9.6
Female Ph.D.s Time to the Doctorate	8.1	0.9	7.1	9.9
Percent with B.A. from Top 20 NRC	15.0	1.3	12.6	17.4
Percent with B.A. from Top 40 NRC	24.0	1.3	21.4	26.5
Percent of Ph.D.s at Top 20 NRC	20.6	5.3	13.4	31.1
Percent of Ph.D.s at Top 40 NRC	37.2	6.6	28.5	51.1
Black Ph.D.s Time to the Doctorate	8.3	1.0	6.6	9.9
White Ph.D.s Time to the Doctorate	8.0	1.0	6.7	9.8
Percent with Primary Support Personal**	21.3	21.7	0.0	50.7
Percent with Primary Support Own Earning**	26.0	3.1	21.7	30.7
Percent of Black Ph.D.s***	3.7	0.5	2.3	4.3
Percent of Hispanic Ph.D.s***	2.0	0.7	0.8	3.2
Percent with Bacc from Research I	33.1	1.3	30.3	35.5
Percent with Bacc from Research I & II	42.9	1.3	40.3	45.0
Percent of Ph.D.s at Research I	52.4	7.1	42.9	66.0
Percent of Ph.D.s at Research I & II	69.6	7.8	59.5	84.0

NOTES: Top 20 and Top 40 NRC refer to the National Research Council's 1981-82 Assessment of Research-Doctorate Programs in the United States. Research I and Research II are obtained from the Carnegie Classification of Colleges and Universities.

*Data from 1964-1983 **Data from 1977-1986 ***Data from 1974-1986

Field Table 4.10: Economics

Variable	Mean of Years	Standard Deviation	Minimum	Maximum
Number of Ph.D.s	601.2	68.9	499.0	729.0
Ph.D.s per 100,000 of U.S. Population	10.3	4.6	5.1	17.7
Total Time to the Doctorate	9.1	0.4	8.5	9.7
Registered Time to the Doctorate	6.3	0.6	5.1	7.2
Percent with Federal Support	31.9	11.5	11.6	46.7
Percent with Private Support	8.9	4.3	3.5	21.1
Percent with T.A.	62.4	7.6	47.5	78.6
Percent with R.A.	43.3	5.2	34.0	51.9
Percent with Spousal Support	28.6	5.9	20.3	39.2
Percent Married	71.7	9.4	57.4	83.4
Mean Number of Dependents	1.4	0.4	0.8	2.1
Percent with Foreign Baccalaureate	8.5	1.9	5.1	13.5
Median Age Starting Ph.D.	22.6	0.2	22.3	23.0
Percent with Definite Employment/Study*	88.7	3.5	81.0	94.1
Percent Negotiating Employment/Study*	5.5	1.7	3.2	9.7
Percent seeking Employment/Study*	5.5	2.2	1.4	9.3
Percent with Temporary Visas	24.2	5.4	16.9	35.9
Percent with Postdoc Study Plan*	2.8	0.8	1.2	4.5
Percent with Same B.S. Fields	60.0	4.2	50.9	65.1
Percent of Women Ph.D.s	11.9	5.1	4.8	22.9

Male Ph.D.s Time to the Doctorate	7.8	0.4	7.0	8.4
Female Ph.D.s Time to the Doctorate	8.2	0.9	6.7	11.5
Percent with B.A. from Top 20 NRC	17.5	1.2	15.7	19.9
Percent with B.A. from Top 40 NRC	26.4	1.6	23.8	29.7
Percent of Ph.D.s at Top 20 NRC	40.1	2.3	35.8	44.5
Percent of Ph.D.s at Top 40 NRC	59.9	2.7	55.8	65.6
Black Ph.D.s Time to the Doctorate	9.0	1.2	7.0	11.5
White Ph.D.s Time to the Doctorate	7.8	0.2	7.4	8.2
Percent with Primary Support Personal**	14.1	14.3	0.0	36.7
Percent with Primary Support Own Earning**	17.5	3.0	14.8	23.5
Percent of Black Ph.D.s***	2.5	1.0	1.1	4.0
Percent of Hispanic Ph.D.s***	1.7	0.9	0.3	2.9
Percent with Bacc from Research I	38.4	2.2	34.4	42.5
Percent with Bacc from Research I & II	49.6	2.7	44.6	56.9
Percent of Ph.D.s at Research I	72.1	2.8	68.2	79.1
Percent of Ph.D.s at Research I & II	87.7	2.1	84.6	92.8

NOTES: Top 20 and Top 40 NRC refer to the National Research Council's 1981-82 Assessment of Research-Doctorate Programs in the United States. Research I and Research II are obtained from the Carnegie Classification of Colleges and Universities.

*Data from 1964-1983 **Data from 1977-1986 ***Data from 1974-1986

Field Table 4.11: Other Social Sciences

Variable	Mean of Years	Standard Deviation	Minimum	Maximum
Number of Ph.D.s	1688.0	327.7	957.0	2124.0
Ph.D.s per 100,000 of U.S. Population	27.8	10.3	14.3	47.8
Total Time to the Doctorate	10.9	0.9	9.9	12.9
Registered Time to the Doctorate	7.2	0.9	5.6	8.8
Percent with Federal Support	41.4	10.4	25.4	54.7
Percent with Private Support	10.1	3.1	6.5	19.5
Percent with T.A.	54.2	3.9	45.4	63.0
Percent with R.A.	36.0	2.8	32.2	41.6
Percent with Spousal Support	32.2	5.3	24.1	41.5
Percent Married	70.4	7.9	59.6	81.1
Mean Number of Dependents	1.4	0.4	0.9	2.0
Percent with Foreign Baccalaureate	5.0	0.9	3.2	6.8
Median Age Starting Ph.D.	22.9	0.2	22.6	23.3
Percent with Definite Employment/Study*	79.5	8.8	63.9	91.0
Percent Negotiating Employment/Study*	7.1	1.6	4.9	9.6
Percent seeking Employment/Study*	12.9	7.4	3.3	27.2
Percent with Temporary Visas	12.3	2.7	9.0	18.2
Percent with Postdoc Study Plan*	6.1	2.6	3.0	10.0
Percent with Same B.S. Fields	40.9	12.4	24.2	55.3
Percent of Women Ph.D.s	27.0	10.2	13.4	44.8

Male Ph.D.s Time to the Doctorate	9.3	1.0	8.2	11.6
Female Ph.D.s Time to the Doctorate	10.1	1.0	8.8	12.2
Percent with B.A. from Top 20 NRC	19.1	1.8	16.2	23.4
Percent with B.A. from Top 40 NRC	26.4	1.6	24.4	30.6
Percent of Ph.D.s at Top 20 NRC	38.9	4.5	32.6	47.6
Percent of Ph.D.s at Top 40 NRC	56.1	2.7	51.9	61.7
Black Ph.D.s Time to the Doctorate	10.0	1.3	7.7	12.2
White Ph.D.s Time to the Doctorate	9.8	1.0	8.6	11.5
Percent with Primary Support Personal**	18.6	19.0	0.0	45.9
Percent with Primary Support Own Earning**	23.5	3.6	18.4	30.0
Percent of Black Ph.D.s***	4.8	0.8	3.2	6.2
Percent of Hispanic Ph.D.s***	2.5	0.9	1.0	4.0
Percent with Bacc from Research I	36.2	2.1	34.1	41.8
Percent with Bacc from Research I & II	46.5	2.1	43.5	51.9
Percent of Ph.D.s at Research I	69.8	4.1	63.0	76.7
Percent of Ph.D.s at Research I & II	86.6	3.0	81.9	92.2

NOTES: Top 20 and Top 40 NRC refer to the National Research Council's 1981-82 Assessment of Research-Doctorate Programs in the United States. Research I and Research II are obtained from the Carnegie Classification of Colleges and Universities.

*Data from 1964-1983 **Data from 1977-1986 ***Data from 1974-1986

Field Table 4.12: Common Variables

Variable	Mean of Years	Standard Deviation	Minimum	Maximum
FACULTY	413700.0	46506.0	299000.0	457000.0
PH.D.S IN TOTAL 11 FIELDS	14906.3	1355.2	11218.0	17263.0
TUITION	515.9	41.9	406.2	568.2
EXPENDST*	57502.8	23629.3	0.0	89951.0
R&D	3221049.2	692787.9	1301242.0	6538280.0
UNEMP	6.7	2.1	3.1	10.9
UNEMP4YR	2.3	0.8	0.9	3.8

NOTE: Acronyms are defined in Appendix B.
*Data from 1971-1986

TABLE A5: Linear Common Variables Model, Mean TTD

Variable	Chemistry	P&A	EAM	Math	Engrg.	Agri. Sci.
WOMEN	0.017	0.014	-0.023	-0.002	0.146	0.037
	(2.40)	(0.06)	(0.46)	(0.00)	(5.57)	(0.59)
AGE	3.741	1.439	0.826	6.173	0.509	1.303
	(37.50)	(3.82)	(2.75)	(6.36)	(0.64)	(4.39)
SUPFED	0.008	0.015	-0.003	-0.010	0.038	0.014
	(2.99)	(2.06)	(0.06)	(0.21)	(2.43)	(0.66)
SUPTA	0.017	-0.015	-0.029	0.006	-0.037	-0.020
	(2.76)	(0.18)	(2.20)	(0.02)	(1.27)	(0.13)
SUPRA	0.004	-0.002	-0.054	-0.048	0.028	0.108
	(0.26)	(0.01)	(9.50)	(2.42)	(0.63)	(1.22)
FORBACC	-0.001	0.017	0.026	0.083	0.044	0.061
	(0.00)	(0.34)	(0.44)	(1.20)	(2.36)	(1.51)
BCARN1ST	-0.040	-0.019	-0.045	0.043	0.026	0.035
	(28.61)	(0.52)	(3.59)	(1.23)	(1.35)	(0.88)
PCARN1ST	-0.011	0.002	-0.023	0.028	-0.026	0.008
	(0.88)	(0.01)	(0.43)	0.14)	(0.47)	(0.03)
FACULTY	-0.000	0.000	0.000	0.000	-0.000	-0.000
	(4.07)	(0.02)	(0.07)	(0.37)	(2.95)	(0.11)
SALRAT1	0.962	-0.851	-2.136	1.240	0.171	1.054
	(7.23)	(0.99)	(6.10)	(0.34)	(0.03)	(0.97)
UNEMP4YR	-0.084	0.081	-0.164	-0.308	-0.070	0.121
	(7.34)	(0.70)	(1.51)	(1.53)	(1.00)	(0.33)
PERPOP	-0.018	-0.032	-0.211	-0.036	-0.028	0.059
	(16.60)	(3.50)	(4.07)	(0.38)	(10.27)	(0.83)
CONSTANT	-76.92	-21.26	3.61	-133.87	-0.365	-33.99
R^2	0.97	0.89	0.78	0.92	0.90	0.19
F	53.42	14.16	6.54	19.04	14.81	1.36
SE	0.07	0.18	0.29	0.34	0.19	0.45
DW	2.51	2.69	2.07	2.22	3.19	1.57

NOTES: (1) On critical F values: For partial F test, F(1, 7, .05) = 5.59; for pp. 175-177

Variable	Bio.Sci.	Health	Psych.	Economics	Soc.Sci.
WOMEN	0.025	0.044	0.047	0.030	0.117
	(2.25)	(1.30)	(2.80)	(0.70)	(8.66)
AGE	1.238	1.380	2.132	0.531	2.550
	(9.38)	(6.80)	(36.91)	(1.36)	(11.14)
SUPFED	0.007	-0.021	0.016	-0.020	0.023
	(0.44)	(0.36)	(3.24)	(2.19)	(0.62)
SUPTA	-0.004	-0.006	-0.034	-0.014	-0.061
	(0.01)	(0.01)	(5.77)	(0.53)	(1.62)
SUPRA	0.024	-0.064	0.026	0.001	-0.007
	(0.95)	(0.77)	(6.03)	(0.01)	(0.04)
FORBACC	0.107	-0.106	-0.105	0.008	0.129
	(3.05)	(1.60)	(0.87)	(0.06)	(2.29)
BCARN1ST	0.005	-0.021	-0.062	0.032	-0.017
	(0.03)	(0.11)	(6.30)	(1.12)	(0.16)
PCARN1ST	0.031	0.007	0.021	-0.010	-0.016
	(1.59)	(0.02)	(1.70)	(0.08)	(0.11)
FACULTY	-0.000	-0.000	0.000	0.000	-0.000
	(3.05)	(1.73)	(0.02)	(0.63)	(0.73)
SALRAT1	0.061	0.521	0.157	0.050	0.720
	(0.01)	(0.10)	(0.28)	(0.00)	(0.62)
UNEMP4YR	-0.058	-0.093	-0.067	-0.076	-0.158
	(0.98)	(0.08)	(1.82)	(0.49)	(1.39)
PERPOP	-0.021	-0.192	-0.036	0.030	0.019
	(11.06)	(3.58)	(7.88)	(0.41)	(0.76)
CONSTANT	-21.20	-15.54	-38.83	-3.96	-46.26
R^2	0.95	0.87	0.99	0.85	0.96
F	28.90	11.46	347.52	9.78	36.55
SE	0.14	0.60	0.10	0.25	0.30
DW	2.88	2.38	2.55	2.26	2.89

overall F test, F(12, 7, .05) = 3.57. (2) Acronyms are defined in Appendix B,

TABLE A5.1: Log Common Variables Model, Mean TTD

Variable	Chemistry	P&A	EAM	Math	Engrg.	Agri. Sci.
WOMEN	0.058	0.008	-0.052	0.040	0.018	-0.033
	(3.54)	(0.07)	(2.16)	(0.22)	(0.39)	(0.44)
AGE	10.944	4.968	2.690	12.414	2.431	4.039
	(21.86)	(8.82)	(5.40)	(6.27)	(1.46)	(5.46)
SUPFED	0.058	0.079	-0.023	-0.027	0.007	-0.047
	(3.65)	(4.52)	(0.31)	(0.14)	(0.02)	(1.16)
SUPTA	0.290	-0.282	-0.093	0.337	0.061	-0.082
	(3.81)	(1.11)	(1.01)	(0.86)	(0.15)	(0.31)
SUPRA	0.159	-0.151	-0.215	-0.214	-0.045	1.066
	(2.47)	(0.41)	(4.96)	(1.84)	(0.03)	(1.61)
FORBACC	0.012	0.019	0.021	0.098	0.054	0.060
	(0.78)	(1.11)	(1.07)	(2.63)	(0.71)	(1.90)
BCARN1ST	-0.150	-0.335	-0.160	0.142	0.111	-0.018
	(11.49)	(5.31)	(2.89)	(0.31)	(0.47)	(0.01)
PCARN1ST	-0.226	-0.102	-0.388	0.117	-0.475	0.049
	(2.57)	(0.20)	(2.41)	(0.03)	(1.50)	(0.02)
FACULTY	-0.243	-0.015	0.172	-0.092	-0.184	0.243
	(12.53)	(0.02)	(0.77)	(0.15)	(0.80)	(0.43)
SALRAT1	0.052	-0.145	-0.264	0.232	-0.078	0.096
	(0.23)	(1.36)	(5.18)	(0.29)	(0.17)	(0.37)
UNEMP4YR	-0.011	0.018	-0.045	-0.065	0.008	-0.031
	(0.75)	(0.74)	(4.52)	(1.45)	(0.16)	(0.49)
PERPOP	-0.028	-0.110	-0.202	-0.098	-0.046	0.059
	(1.53)	(6.55)	(7.11)	(0.38)	(0.71)	(0.41)
CONSTANT	-29.68	-10.86	-4.19	-36.88	-1.48	-18.25
R^2	0.95	0.93	0.84	0.92	0.85	0.25
F	29.38	22.46	9.18	18.53	9.85	1.54
SE	0.01	0.02	0.03	0.04	0.03	0.05
DW	2.85	3.02	2.51	2.22	3.00	1.85

NOTES: (1) On critical F values: For partial F test, F(1, 7, .05) = 5.59; for pp. 175-177

Variable	Bio.Sci.	Health	Psych.	Economics	Soc.Sci.
WOMEN	0.133	0.055	0.215	0.012	0.162
	(9.91)	(0.17)	(2.69)	(0.07)	(1.28)
AGE	3.254	3.49	6.422	1.181	5.980
	(11.25)	(14.91)	(37.34)	(0.85)	(7.35)
SUPFED	0.054	-0.174	0.136	-0.072	-0.066
	(0.61)	(1.31)	(4.94)	(2.88)	(0.18)
SUPTA	0.046	0.026	-0.245	-0.149	-0.256
	(0.08)	(0.02)	(6.87)	(1.46)	(0.76)
SUPRA	0.088	-0.108	0.146	0.038	0.037
	(0.80)	(0.49)	(4.21)	(0.27)	(0.07)
FORBACC	0.059	-0.098	-0.009	0.016	0.075
	(4.18)	(2.27)	(0.09)	(0.30)	(2.56)
BCARN1ST	0.015	-0.176	-0.137	0.147	-0.084
	(0.02)	(0.71)	(1.32)	(1.69)	(0.21)
PCARN1ST	0.243	0.064	0.047	-0.002	-0.129
	(1.79)	(0.03)	(0.14)	(0.00)	(0.13)
FACULTY	-0.333	-0.193	-0.104	0.193	0.107
	(7.98)	(0.20)	(0.29)	(1.07)	(0.14)
SALRAT1	-0.014	0.067	-0.025	0.026	0.082
	(0.01)	(0.06)	(0.16)	(0.05)	(0.23)
UNEMP4YR	-0.022	-0.012	0.004	-0.019	-0.028
	(1.81)	(0.03)	(0.05)	(0.85)	(0.59)
PERPOP	-0.135	-0.169	-0.137	0.020	0.113
	(8.01)	(3.82)	(2.32)	(0.12)	(1.27)
CONSTANT	-5.50	-4.50	-16.56	-3.89	-16.76
R^2	0.95	0.83	0.99	0.87	0.92
F	34.28	8.94	209.58	11.25	20.08
SE	0.02	0.06	0.01	0.03	0.04
DW	3.08	2.34	2.16	2.50	2.47

overall F test, F(12, 7, .05) = 3.57. (2) Acronyms are defined in Appendix B,

TABLE A5.2: Unique Variables Model, Mean TTD

Variable	Chemistry	P&A	EAM	Math	Engrg.	Agri. Sci.
MARRIED						
WOMEN			0.052			
			(6.62)			
TEMP						
DEPEND	0.387					
	(14.44)					
AGE	3.544	2.133		4.461	1.519	1.089
	(79.56)	(24.97)		(49.12)	(31.58)	(36.45)
SAMEFLD				-0.034		
				(8.42)		
SUPFED						-0.012
						(7.52)
SUPRA			-0.065			
			(16.32)			
SUPTA	0.028	0.044		0.045		
	(17.54)	(16.61)		(11.73)		
TUITION						0.006
						(28.68)
FORBACC						
BCARN1ST	-0.036		-0.069			
	(17.38)		(13.48)			
PCARN1ST						
SDRSAL1O						
SALRAT1						1.632
						(12.47)
SEEK		0.018				
		(6.62)				
PERPOP			-0.191		0.014	
			(6.38)		(126.37)	
CONSTANT	-73.730	-42.640	17.096	-91.761	-25.271	-21.575
R^2	0.92	0.90	0.67	0.92	0.88	0.70
F	50.26	56.02	10.67	71.92	70.17	12.31
SE	0.07	0.12	0.24	0.23	0.13	0.19
DW	2.58	2.59	1.96	1.81	2.34	1.69

NOTES: (1) On critical F values: For 2-variable model, partial F is F(1, 17, .05) 16, .05) = 4.49; overall F is F(3, 16, .05) = 3.24. For 4-variable model, partial F defined in Appendix B, pp. 175-177.

Variable	Bio.Sci.	Health	Psych.	Economics	Soc.Sci.
MARRIED			0.081 (25.92)		
WOMEN					
TEMP					0.137 (8.16)
DEPEND					
AGE	1.930 (77.54)	2.016 (57.14)		0.906 (19.11)	1.316 (18.99)
SAMEFLD					
SUPFED			-0.092 (92.27)		
SUPRA					
SUPTA					
TUITION					
FORBACC		-0.172 (6.25)			
BCARN1ST			-0.109 (8.57)	0.044 (7.21)	
PCARN1ST	0.052 (12.06)				
SDRSAL10				-0.0001 (5.93)	-0.0001 (6.25)
SALRAT1			1.629 (5.24)		
SEEK					
PERPOP	-0.011 (22.94)	-0.212 (9.79)		-0.046 (35.98)	
CONSTANT	-38.225	-33.575	9.311	-10.537	-19.153
R^2	0.91	0.81	0.95	0.84	0.97
F	61.66	27.58	100.94	26.14	176.67
SE	0.12	0.48	0.22	0.17	0.18
DW	1.97	1.61	1.88	1.94	1.91

= 4.45; overall F is F(2, 17, .05) = 3.59. For 3-variable model, partial F is F(1, is F(1, 15, .05) = 4.54; overall F is F(4, 15, .05) = 3.06. (2) Acronyms are

TABLE A6: Linear Common Variables Model, Mean RTD

Variable	Chemistry	P&A	EAM	Math	Engrg.	Agri. Sci.
WOMEN	0.009	0.014	-0.020	0.050	0.101	0.034
	(0.56)	(0.10)	(0.85)	(1.25)	(5.34)	(2.74)
AGE	2.086	0.640	0.946	1.131	0.158	0.394
	(10.57)	(1.29)	(8.90)	(0.88)	(0.12)	(2.14)
SUPFED	0.001	0.005	-0.022	-0.004	0.031	0.007
	(0.08)	(0.39)	(6.31)	(0.15)	(3.10)	(0.98)
SUPTA	0.002	-0.016	-0.016	0.017	-0.039	0.014
	(0.02)	(0.34)	(1.65)	(0.49)	(2.77)	(0.31)
SUPRA	0.007	0.009	-0.016	-0.015	0.032	0.013
	(0.89)	(0.22)	(2.08)	(0.94)	(1.55)	(0.10)
FORBACC	0.021	0.008	0.046	0.059	0.025	0.023
	(3.13)	(0.14)	(3.25)	(2.56)	(1.56)	(1.12)
BCARN1ST	-0.023	-0.021	-0.016	0.011	0.021	0.049
	(8.70)	(1.09)	(1.10)	(0.34)	(1.89)	(9.35)
PCARN1ST	-0.006	-0.008	0.016	0.017	-0.006	-0.044
	(0.24)	(0.14)	(0.51)	(0.21)	(0.06)	(5.21)
FACULTY	-0.000	0.000	0.000	-0.000	-0.000	0.000
	(0.08)	(1.12)	(8.55)	(0.08)	(0.34)	(3.47)
SALRAT1	0.182	-0.605	-0.949	-1.732	0.316	0.246
	(0.24)	(0.85)	(2.97)	(2.74)	(0.23)	(0.28)
UNEMP4YR	-0.096	-0.026	-0.238	0.021	-0.009	0.064
	(8.79)	(0.12)	(7.86)	(0.03)	(0.03)	(0.49)
PERPOP	-0.008	-0.022	-0.132	0.036	-0.014	0.035
	(2.92)	(2.91)	(3.96)	(1.53)	(4.75)	(1.54)
CONSTANT	-40.55	-5.31	-13.39	-19.47	0.61	-7.53
R^2	0.97	0.92	0.95	0.96	0.93	0.89
F	44.76	19.56	31.03	39.50	20.84	13.35
SE	0.07	0.14	0.18	0.17	0.13	0.20
DW	2.62	3.20	2.13	2.42	3.08	2.87

NOTES: (1) On critical F values: For partial F test, F(1, 7, .05) = 5.59; for pp. 175-177

Variable	Bio.Sci.	Health	Psych.	Economics	Soc.Sci.
WOMEN	0.013	0.001	0.069	-0.004	0.142
	(3.80)	(0.00)	(1.98)	(0.12)	(15.05)
AGE	1.166	0.020	0.610	-0.523	1.765
	(56.84)	(0.65)	(0.98)	(1.28)	(6.27)
SUPFED	-0.015	-0.020	0.008	-0.032	0.064
	(12.96)	(1.51)	(0.29)	(5.12)	(5.58)
SUPTA	0.045	-0.008	-0.032	-0.020	-0.084
	(12.07)	(0.07)	(1.66)	(1.11)	(3.64)
SUPRA	-0.010	0.050	0.042	-0.029	-0.003
	(1.01)	(2.11)	(4.87)	(3.09)	(0.01)
FORBACC	-0.031	-0.028	0.069	-0.048	0.287
	(1.74)	(0.50)	(0.12)	(2.28)	(13.38)
BCARN1ST	0.004	0.014	-0.0005	-0.005	-0.005
	(0.14)	(0.22)	(0.00)	(0.22)	(0.01)
PCARN1ST	-0.012	-0.024	-0.009	-0.038	-0.047
	(1.62)	(1.15)	(0.10)	(1.21)	(1.17)
FACULTY	-0.000	-0.000	-0.000	0.000	0.000
	(8.54)	(1.10)	(0.34)	(4.53)	(0.00)
SALRAT1	-0.132	-0.137	0.178	0.491	0.834
	(0.30)	(0.03)	(0.12)	(0.39)	(0.98)
UNEMP4YR	-0.045	-0.092	0.057	-0.089	-0.341
	(3.92)	(0.38)	(0.42)	(0.66)	(7.54)
PERPOP	-0.002	-0.010	0.012	0.031	-0.010
	(0.44)	(0.05)	(0.27)	(0.42)	(0.27)
CONSTANT	-18.68	4.57	-9.53	17.08	-33.23
R^2	0.99	0.85	0.96	0.94	0.97
F	165.42	9.81	42.75	25.63	46.27
SE	0.05	0.28	0.18	0.25	0.28
DW	3.44	2.42	2.83	2.81	2.49

overall F test, F(12, 7, .05) = 3.57. (2) Acronyms are defined in Appendix B,

TABLE A6.1: Log Common Variables Model, Mean RTD

Variable	Chemistry	P&A	EAM	Math	Engrg.	Agri. Sci.
WOMEN	0.030	-0.003	-0.050	0.053	0.027	0.035
	(0.53)	(0.01)	(4.49)	(1.13)	(0.80)	(0.96)
AGE	6.022	2.594	3.300	6.300	1.184	1.950
	(3.80)	(2.13)	(11.53)	(4.56)	(0.34)	(2.41)
SUPFED	0.004	0.014	-0.066	-0.035	0.015	0.008
	(0.01)	(0.13)	(3.77)	(0.65)	(0.10)	(0.06)
SUPTA	0.197	-0.039	-0.144	0.139	-0.040	0.115
	(1.01)	(0.02)	(3.47)	(0.41)	(0.06)	(1.15)
SUPRA	0.113	0.036	0.021	-0.127	0.059	0.157
	(0.71)	(0.02)	(0.07)	(1.82)	(0.06)	(0.07)
FORBACC	0.032	0.015	0.023	0.067	0.041	-0.004
	(3.20)	(0.55)	(1.73)	(3.47)	(0.40)	(0.02)
BCARN1ST	-0.109	-0.240	0.007	0.010	0.168	0.411
	(3.50)	(2.41)	(0.01)	(0.00)	(1.05)	(5.79)
PCARN1ST	-0.213	-0.113	-0.066	0.382	-0.335	-0.681
	(1.31)	(0.21)	(0.10)	(0.91)	(0.72)	(6.85)
FACULTY	-0.098	0.062	0.561	0.036	-0.060	0.429
	(1.18)	(0.32)	(11.59)	(0.06)	(0.08)	(2.52)
SALRAT1	-0.062	-0.113	-0.062	-0.200	0.033	0.024
	(0.18)	(0.73)	(0.40)	(0.61)	(0.03)	(0.04)
UNEMP4YR	-0.023	-0.014	-0.070	-0.029	-0.005	0.017
	(1.99)	(0.42)	(15.64)	(0.84)	(0.05)	(0.28)
PERPOP	-0.012	-0.083	-0.158	0.013	-0.008	0.118
	(0.15)	(3.34)	(6.21)	(0.02)	(0.02)	(3.11)
CONSTANT	-15.84	-5.32	-14.23	-20.02	-2.14	-9.96
R^2	0.94	0.93	0.97	0.97	0.89	0.87
F	23.93	22.82	48.38	45.85	13.94	11.68
SE	0.02	0.02	0.02	0.03	0.03	0.04
DW	2.83	3.05	2.43	2.45	3.31	2.71

NOTES: (1) On critical F values: For partial F test, F(1, 7, .05) = 5.59; for pp. 175-177

Variable	Bio.Sci.	Health	Psych.	Economics	Soc.Sci.
WOMEN	0.056	-0.059	-0.007	-0.011	0.324
	(5.51)	(0.36)	(0.00)	(0.02)	(1.59)
AGE	3.459	0.929	3.448	-1.706	6.398
	(39.29)	(1.92)	(2.06)	(0.80)	(2.64)
SUPFED	-0.113	-0.237	-0.037	-0.092	0.219
	(8.25)	(4.39)	(0.07)	(2.10)	(0.61)
SUPTA	0.329	0.047	-0.268	0.152	-0.477
	(12.92)	(0.12)	(1.58)	(0.68)	(0.83)
SUPRA	-0.045	0.183	0.135	-0.177	0.093
	(0.66)	(2.56)	(0.69)	(2.57)	(0.14)
FORBACC	-0.019	-0.038	0.028	0.062	0.198
	(1.28)	(0.63)	(0.19)	(1.93)	(5.61)
BCARN1ST	0.009	0.040	-0.216	0.013	-0.163
	(0.01)	(0.07)	(0.63)	(0.01)	(0.24)
PCARN1ST	-0.095	-0.159	-0.142	-0.156	-0.480
	(0.85)	(0.37)	(0.24)	(0.24)	(0.57)
FACULTY	-0.221	-0.017	0.177	0.500	0.236
	(10.89)	(0.00)	(0.16)	(3.23)	(0.22)
SALRAT1	-0.041	-0.030	0.034	0.091	0.116
	(0.30)	(0.02)	(0.06)	(0.26)	(0.14)
UNEMP4YR	-0.018	-0.019	-0.019	-0.037	-0.099
	(3.86)	(0.14)	(0.18)	(1.39)	(2.33)
PERPOP	-0.043	0.006	-0.065	-0.038	-0.032
	(2.53)	(0.01)	(0.10)	(0.20)	(0.03)
CONSTANT	-6.32	-0.23	-8.95	1.63	-18.96
R^2	0.99	0.84	0.95	0.95	0.91
F	156.26	9.41	30.62	29.81	16.59
SE	0.01	0.04	0.03	0.04	0.06
DW	3.42	2.35	2.48	2.88	2.43

overall F test, F(12, 7, .05) = 3.57. (2) Acronyms are defined in Appendix B,

TABLE A6.2: Unique Variables Model, Mean RTD

Variable	Chemistry	P&A	EAM	Math	Engrg.	Agri. Sci.
MARRIED		-0.026	-0.037			
		(27.21)	(94.81)			
WOMEN				0.049		
				(7.22)		
TEMP			0.034			
			(6.21)			
AGE	1.516					
	(12.26)					
SAMEFLD					-0.033	
					(101.94)	
SUPFED						
SUPRA						
SUPPRIV						
SUPTA		0.024				0.048
		(5.59)				(5.68)
FORBACC	0.033				0.063	0.023
	(17.77)				(11.15)	(7.86)
BCARN1ST			-0.082			
			(17.01)			
PCARN1ST		-0.049				
		(11.47)				
BTOP20			0.090			
			(6.64)			
SDRSAL10						-0.0002
						(24.63)
SALRAT1	-0.563			-0.867		
	(15.42)			(5.54)		
SALRATIO						
SEEK						
DEFIN					-0.013	-0.016
					(5.73)	(4.69)
CONSTANT	27.249	10.287	10.652	6.52	9.537	8.238
R^2	0.91	0.91	0.89	0.97	0.93	0.82
F	67.36	63.47	40.82	183.78	84.03	30.36
SE	0.07	0.10	0.18	0.10	0.09	0.16
DW	1.97	2.42	2.12	1.94	2.33	1.11

NOTES: (1) On critical F values: For 2-variable model, partial F is F(1, 17, .05) 16, .05) = 4.49; overall F is F(3, 16, .05) = 3.24. For 4-variable model, partial F defined in Appendix B, pp. 175-177.

Variable	Bio.Sci.	Health	Psych.	Economics	Soc.Sci.
MARRIED					
WOMEN					
TEMP		0.054 (13.78)	0.262 (9.15)	0.057 (54.69)	0.172 (90.86)
AGE					
SAMEFLD					
SUPFED		-0.025 (32.58)	-0.027 (20.61)		
SUPRA	0.025 (13.97)				
SUPPRIV				-0.070 (76.86)	-0.084 (139.26)
SUPTA					
FORBACC				0.066 (11.65)	
BARN1ST					
PCARN1ST					
BTOP20					
SDRSAL10	-0.0001 (11.33)		-0.0002 (10.87)		-0.0001 (37.38)
SALRATI					
SALRATIO		-0.756 (8.53)			
SEEK	0.023 (10.43)				
DEFIN					
CONSTANT	6.422	8.533	9.938	4.967	8.134
R^2	0.95	0.85	0.96	0.95	0.99
F	112.41	36.08	146.70	134.25	951.06
SE	0.09	0.19	0.13	0.14	0.08
DW	1.38	2.27	1.88	1.74	2.24

= 4.45; overall F is F(2, 17, .05) = 3.59. For 3-variable model, partial F is F(1, is F(1, 15, .05) = 4.54; overall F is F(4, 15, .05) = 3.06. (2) Acronyms are

TABLE A7: Linear Common Variables Model, Mean TPGE

Variable	Chemistry	P&A	EAM	Math	Engrg.	Agri. Sci.
WOMEN	0.005	0.010	0.001	-0.029	0.012	-0.006
	(0.47)	(0.10)	(0.03)	(2.63)	(0.63)	(0.17)
AGE	1.432	0.307	0.256	1.884	0.522	0.243
	(10.99)	(1.29)	(0.87)	(15.18)	(10.95)	(1.50)
SUPFED	-0.004	0.006	0.005	-0.003	0.004	0.006
	(1.23)	(0.39)	(0.37)	(0.56)	(0.36)	(1.27)
SUPTA	0.016	-0.015	-0.009	-0.012	0.003	-0.008
	(4.98)	(0.34)	(0.69)	(1.52)	(0.10)	(0.19)
SUPRA	0.015	0.005	0.000	-0.012	-0.009	0.029
	(7.72)	(0.22)	(0.17)	(3.64)	(1.04)	(0.89)
FORBACC	-0.013	-0.002	0.036	0.040	0.005	0.001
	(2.40)	(0.14)	(2.76)	(7.09)	(0.50)	(0.00)
BCARN1ST	0.002	-0.002	-0.001	0.016	0.010	-0.006
	(0.09)	(1.09)	(0.01)	(4.09)	(3.13)	(0.27)
PCARN1ST	-0.001	-0.000	0.017	0.006	0.004	0.002
	(0.25)	(0.14)	(0.74)	(0.15)	(0.16)	(0.02)
FACULTY	0.000	0.000	-0.000	0.000	-0.000	-0.000
	(0.11)	(1.12)	(0.03)	(5.56)	(0.05)	(1.93)
SALRAT1	0.424	-0.414	0.272	1.081	-0.107	0.312
	(2.80)	(0.85)	(0.32)	(6.63)	(0.22)	(0.84)
UNEMP4YR	0.015	0.048	0.087	-0.021	-0.028	0.022
	(0.44)	(0.12)	(1.41)	(17.46)	(2.59)	(0.11)
PERPOP	-0.009	-0.002	-0.032	-0.033	-0.005	-0.015
	(0.09)	(2.91)	(0.31)	(8.17)	(4.36)	(0.52)
CONSTANT	-32.29	-5.50	-6.26	-42.89	-10.98	-5.49
R^2	0.76	0.89	0.05	0.94	0.84	0.52
F	5.93	14.02	1.09	26.30	9.04	2.73
SE	0.05	0.02	0.16	0.07	0.05	0.14
DW	2.63	2.66	2.54	2.28	2.79	1.92

NOTES: (1) On critical F values: For partial F test, F(1, 7, .05) = 5.59; for pp. 175-177

Variable	Bio.Sci.	Health	Psych.	Economics	Soc.Sci.
WOMEN	0.011	0.020	0.009	0.034	0.012
	(3.82)	(2.81)	(0.31)	(1.63)	(1.57)
AGE	0.502	0.524	0.460	0.589	0.937
	(12.61)	(10.08)	(4.88)	(3.02)	(24.99)
SUPFED	0.002	0.016	0.005	0.010	0.010
	(0.29)	(2.27)	(0.92)	(0.94)	(1.87)
SUPTA	-0.007	-0.014	0.004	-0.014	-0.031
	(0.37)	(0.56)	(0.24)	(0.98)	(7.16)
SUPRA	-0.004	-0.036	-0.009	0.002	0.001
	(0.21)	(2.54)	(2.24)	(0.03)	(0.02)
FORBACC	0.014	-0.003	-0.004	-0.009	0.013
	(0.40)	(0.02)	(0.00)	(0.16)	(0.39)
BCARN1ST	0.012	0.014	0.006	0.001	-0.017
	(1.34)	(0.49)	(0.18)	(0.00)	(2.50)
PCARN1ST	0.003	0.001	-0.008	0.047	-0.017
	(0.12)	(0.00)	(0.60)	(3.21)	(2.08)
FACULTY	-0.000	0.000	-0.000	0.000	-0.000
	(1.59)	(0.85)	(0.90)	(0.34)	(2.17)
SALRAT1	0.121	0.521	0.054	0.151	0.379
	(0.31)	(1.03)	(0.09)	(0.07)	(2.86)
UNEMP4YR	-0.006	0.002	0.008	-0.052	0.043
	(0.10)	(0.00)	(0.07)	(0.42)	(1.68)
PERPOP	-0.004	-0.040	-0.006	-0.003	0.003
	(3.59)	(1.56)	(0.72)	(0.01)	(0.34)
CONSTANT	-10.50	-13.03	-8.65	-16.51	-17.58
R^2	0.93	0.84	0.97	0.51	0.94
F	21.76	9.01	57.18	2.64	24.24
SE	0.05	0.19	0.06	0.18	0.07
DW	2.46	3.17	2.13	1.98	2.83

overall F test, F(12, 7, .05) = 3.57. (2) Acronyms are defined in Appendix B,

TABLE A7.1: Log Common Variables Model, Mean TPGE

Variable	Chemistry	P&A	EAM	Math	Engrg.	Agri. Sci.
WOMEN	0.126	0.148	-0.042	-0.041	-0.055	-0.126
	(0.38)	(1.48)	(0.04)	(0.03)	(1.11)	(1.36)
AGE	60.399	19.341	8.323	28.376	15.048	4.413
	(15.09)	(8.47)	(0.97)	(3.60)	(17.55)	(1.08)
SUPFED	-0.038	0.321	0.029	0.125	-0.019	0.027
	(0.04)	(4.74)	(0.01)	(0.33)	(0.05)	(0.06)
SUPTA	2.249	-2.704	-0.539	0.774	0.608	-0.407
	(5.19)	(6.46)	(0.64)	(0.50)	(4.72)	(1.26)
SUPRA	-1.613	1.702	0.012	-0.616	-0.832	2.594
	(5.73)	(3.29)	(0.00)	(1.68)	(3.61)	(1.57)
FORBACC	-0.139	-0.039	0.238	0.676	0.006	0.180
	(2.37)	(0.29)	(2.54)	(13.87)	(0.00)	(2.81)
BCARN1ST	0.135	-0.910	-0.043	0.632	0.296	-0.501
	(0.21)	(2.49)	(0.00)	(0.68)	(1.05)	(0.75)
PCARN1ST	0.136	-0.720	1.190	-0.478	-0.290	-0.135
	(0.02)	(0.62)	(0.42)	(0.06)	(0.18)	(0.02)
FACULTY	-0.204	0.576	0.264	-0.202	0.089	-0.661
	(0.20)	(2.00)	(0.03)	(0.08)	(0.06)	(0.52)
SALRAT1	1.017	-1.242	0.166	2.570	-0.464	0.246
	(1.97)	(6.32)	(0.04)	(3.92)	(1.92)	(0.39)
UNEMP4YR	0.046	0.242	0.205	-0.536	-0.023	-0.126
	(0.31)	(8.77)	(1.78)	(10.92)	(0.42)	(1.36)
PERPOP	-0.141	0.019	-0.022	-1.087	-0.205	-0.401
	(0.86)	(0.01)	(0.00)	(5.19)	(4.38)	(3.14)
CONSTANT	-189.57	-58.59	-32.99	-86.93	-45.95	-11.92
R^2	0.75	0.80	0.00	0.92	0.83	0.55
F	5.86	7.27	0.91	19.00	8.55	2.92
SE	0.09	0.08	0.20	0.13	0.05	0.12
DW	2.88	2.26	2.79	2.72	2.50	1.97

NOTES: (1) On critical F values: For partial F test, F(1, 7, .05) = 5.59; for pp. 175-177

Variable	Bio.Sci.	Health	Psych.	Economics	Soc.Sci.
WOMEN	0.454	0.469	0.727	0.076	0.218
	(10.55)	(4.11)	(2.58)	(0.04)	(0.51)
AGE	10.700	8.910	11.001	10.979	19.050
	(11.12)	(32.14)	(9.16)	(1.10)	(16.51)
SUPFED	0.071	0.537	0.371	-0.011	0.084
	(0.10)	(4.13)	(3.09)	(0.00)	(0.06)
SUPTA	-0.357	-0.249	-0.092	-0.726	-1.252
	(0.45)	(0.60)	(0.08)	(0.52)	(4.06)
SUPRA	-0.293	-0.519	-0.111	0.331	-0.040
	(0.82)	(3.74)	(0.20)	(0.30)	(0.02)
FORBACC	0.102	-0.057	0.142	0.071	0.113
	(1.14)	(0.25)	(2.10)	(0.08)	(1.29)
BCARN1ST	0.340	0.231	0.694	0.056	-0.325
	(0.68)	(0.41)	(2.85)	(0.00)	(0.67)
PCARN1ST	0.481	-0.195	-0.558	2.359	-0.960
	(0.64)	(0.10)	(1.61)	(1.82)	(1.60)
FACULTY	-0.784	-0.165	-1.234	1.075	-0.152
	(4.05)	(0.05)	(3.46)	(0.50)	(0.06)
SALRAT1	0.277	0.552	0.026	0.732	0.502
	(0.42)	(1.27)	(0.01)	(0.57)	(1.89)
UNEMP4YR	-0.081	-0.049	0.165	-0.091	0.057
	(2.25)	(0.17)	(5.78)	(0.28)	(0.54)
PERPOP	-0.371	-0.348	0.195	0.148	0.269
	(5.50)	(5.34)	(0.39)	(0.10)	(1.61)
CONSTANT	-24.70	-26.37	-22.70	-57.66	-49.44
R^2	0.94	0.88	0.98	0.47	0.90
F	26.49	12.34	72.37	2.39	15.15
SE	0.05	0.10	0.05	0.20	0.08
DW	2.56	3.12	1.80	2.14	2.71

overall F test, F(12, 7, .05) = 3.57. (2) Acronyms are defined in Appendix B,

TABLE A8: Linear Common Variables Model, Mean TNEU

Variable	Chemistry	P&A	EAM	Math	Engrg.	Agri. Sci.
WOMEN	0.007	-0.023	0.037	-0.008	0.050	0.011
	(0.20)	(0.40)	(1.54)	(0.02)	(0.97)	(0.10)
AGE	0.106	0.139	-0.260	2.458	-0.209	0.682
	(0.02)	(0.08)	(0.35)	(2.70)	(0.16)	(2.03)
SUPFED	0.008	0.007	0.012	-0.003	0.012	-0.001
	(1.95)	(0.98)	(0.88)	(0.06)	(0.36)	(0.08)
SUPTA	0.006	0.000	0.002	-0.010	-0.002	-0.019
	(0.23)	(0.00)	(0.01)	(0.10)	(0.00)	(0.18)
SUPRA	0.009	-0.000	-0.036	-0.018	-0.004	0.055
	(0.86)	(0.00)	(5.44)	(0.87)	(0.02)	(0.54)
FORBACC	-0.007	-0.009	-0.038	-0.017	0.029	0.040
	(0.22)	(0.20)	(1.17)	(0.14)	(1.54)	(1.10)
BCARN1ST	-0.016	-0.006	-0.024	0.018	0.006	-0.007
	(2.80)	(0.11)	(1.28)	(0.57)	(0.10)	(0.06)
PCARN1ST	-0.012	-0.002	-0.054	-0.002	-0.006	0.048
	(0.64)	(0.01)	(2.98)	(0.00)	(0.04)	(1.91)
FACULTY	-0.000	-0.000	-0.000	0.000	-0.000	-0.000
	(3.46)	(1.34)	(2.02)	(0.54)	(3.21)	(0.70)
SALRAT1	0.220	-0.208	-1.090	1.644	-0.173	0.255
	(0.22)	(0.13)	(2.01)	(1.60)	(0.05)	(0.10)
UNEMP4YR	0.000	0.085	0.045	-0.090	-0.017	-0.005
	(0.00)	(1.74)	(0.14)	(0.35)	(0.09)	(0.00)
PERPOP	-0.006	-0.008	-0.020	-0.040	-0.014	0.032
	(1.27)	(0.53)	(0.05)	(1.21)	(3.65)	(0.41)
CONSTANT	-1.05	-1.58	18.49	-55.49	8.78	-20.32
R^2	0.65	0.46	0.72	0.22	0.19	0.49
F	3.95	2.37	5.11	1.45	1.36	2.51
SE	0.09	0.12	0.26	0.20	0.16	0.35
DW	2.48	2.21	2.23	2.25	2.79	1.78

NOTES: (1) On critical F values: For partial F test, F(1, 7, .05) = 5.59; for pp. 175-177

Variable	Bio.Sci.	Health	Psych.	Economics	Soc.Sci.
WOMEN	0.002	0.020	-0.009	-0.002	-0.019
	(0.07)	(0.45)	(0.12)	(0.01)	(0.30)
AGE	-0.189	0.823	0.766	0.482	0.208
	(1.09)	(3.88)	(5.23)	(2.15)	(0.10)
SUPFED	0.021	-0.020	0.002	0.001	-0.047
	(17.46)	(0.52)	(0.08)	(0.01)	(3.43)
SUPTA	-0.034	0.007	-0.005	-0.017	0.045
	(4.95)	(0.02)	(0.12)	(1.48)	(1.17)
SUPRA	0.027	-0.072	-0.002	0.017	-0.002
	(5.84)	(1.59)	(0.02)	(2.02)	(0.00)
FORBACC	0.078	-0.087	-0.178	-0.037	-0.162
	(8.21)	(1.75)	(2.73)	(2.70)	(4.75)
BCARN1ST	-0.011	-0.062	-0.070	0.028	0.001
	(0.68)	(1.55)	(8.84)	(1.69)	(0.00)
PCARN1ST	0.028	0.025	-0.040	-0.012	0.052
	(6.21)	(0.45)	(6.53)	(0.24)	(1.57)
FACULTY	-0.000	-0.000	0.000	-0.000	-0.000
	(1.08)	(2.66)	(0.67)	(5.49)	(0.70)
SALRAT1	0.173	0.238	0.090	-0.208	-0.294
	(0.38)	(0.03)	(0.10)	(0.14)	(0.14)
UNEMP4YR	-0.008	0.002	-0.111	0.076	0.134
	(0.08)	(0.00)	(5.38)	(0.94)	(1.30)
PERPOP	-0.014	-0.155	-0.036	-0.001	0.030
	(25.33)	(3.75)	(8.40)	(0.00)	(2.67)
CONSTANT	-3.71	-9.53	-14.31	-5.55	5.40
R^2	0.96	0.62	0.92	0.94	0.80
F	35.67	3.57	18.47	25.23	7.49
SE	0.06	0.47	0.10	0.17	0.26
DW	2.83	2.62	3.17	2.68	2.54

overall F test, F(12, 7, .05) = 3.57. (2) Acronyms are defined in Appendix B,

TABLE A8.1: Log Common Variables Model, Mean TNEU

Variable	Chemistry	P&A	EAM	Math	Engrg.	Agri. Sci.
WOMEN	0.284	0.004	-0.106	0.103	0.078	-0.154
	(1.23)	(0.00)	(0.44)	(0.07)	(0.50)	(0.75)
AGE	19.371	7.379	1.668	27.834	-1.722	10.425
	(0.99)	(0.46)	(0.06)	(1.52)	(0.05)	(2.77)
SUPFED	0.655	0.578	0.025	0.059	0.076	-0.221
	(6.66)	(5.74)	(0.01)	(0.03)	(0.18)	(2.00)
SUPTA	-0.436	-2.216	0.488	0.502	0.159	-0.350
	(0.12)	(1.63)	(0.87)	(0.09)	(0.07)	(0.43)
SUPRA	1.808	1.687	-1.079	-0.442	-0.107	2.208
	(4.58)	(1.21)	(3.91)	(0.38)	(0.01)	(0.53)
FORBACC	-0.042	0.081	-0.009	0.004	0.291	0.201
	(0.14)	(0.46)	(0.01)	(0.00)	(1.44)	(1.61)
BCARN1ST	-0.636	-0.755	-0.527	0.645	0.147	-0.751
	(2.98)	(0.64)	(0.98)	(0.31)	(0.06)	(0.78)
PCARN1ST	-0.984	-0.948	-2.490	-1.105	-0.405	1.803
	(0.70)	(0.41)	(3.09)	(0.13)	(0.08)	(1.93)
FACULTY	-1.609	-0.744	-0.567	-0.169	-1.443	0.284
	(7.90)	(1.24)	(0.26)	(0.02)	(3.40)	(0.04)
SALRAT1	-0.098	-0.247	-1.164	1.301	-0.075	0.100
	(0.01)	(0.09)	(3.15)	(0.44)	(0.01)	(0.03)
UNEMP4YR	0.020	0.237	-0.023	-0.017	0.074	-0.150
	(0.04)	(3.14)	(0.04)	(0.01)	(0.95)	(0.89)
PERPOP	-0.092	-0.373	-0.303	-0.238	-0.062	0.117
	(0.24)	(1.79)	(0.50)	(0.11)	(0.09)	(0.12)
CONSTANT	-41.61	-5.74	19.13	-82.49	24.59	-48.97
R^2	0.75	0.67	0.71	0.00	0.09	0.59
F	5.79	4.28	4.94	0.94	1.16	3.31
SE	0.11	0.12	0.15	0.19	0.10	0.18
DW	2.94	2.69	2.55	1.97	2.52	1.75

NOTES: (1) On critical F values: For partial F test, F(1, 7, .05) = 5.59; for pp. 175-177

Variable	Bio.Sci.	Health	Psych.	Economics	Soc.Sci.
WOMEN	0.211	0.092	0.713	0.159	-0.002
	(2.36)	(0.06)	(0.83)	(0.87)	(0.00)
AGE	3.126	7.531	13.561	4.552	0.896
	(0.97)	(7.95)	(4.69)	(0.96)	(0.02)
SUPFED	0.962	-0.404	0.556	0.052	-0.686
	(18.25)	(0.81)	(2.34)	(0.11)	(2.59)
SUPTA	-0.778	0.034	-0.338	-0.420	0.493
	(2.20)	(0.00)	(0.37)	(0.89)	(0.38)
SUPRA	0.671	-0.555	0.328	0.193	-0.002
	(4.43)	(1.48)	(0.60)	(0.52)	(0.00)
FORBACC	0.309	-0.280	-0.269	-0.185	-0.149
	(10.67)	(2.11)	(2.55)	(2.93)	(1.35)
BCARN1ST	-0.063	-1.028	-0.886	0.689	0.219
	(0.02)	(2.78)	(1.56)	(2.83)	(0.18)
PCARN1ST	1.299	0.513	1.288	-0.346	0.895
	(4.80)	(0.24)	(2.89)	(0.20)	(0.85)
FACULTY	-0.625	-0.757	-0.515	-1.225	-0.381
	(2.64)	(0.36)	(0.20)	(3.29)	(0.24)
SALRAT1	0.045	0.165	-0.117	-0.105	-0.051
	(0.01)	(0.04)	(0.09)	(0.06)	(0.01)
UNEMP4YR	0.044	0.028	-0.034	0.065	0.062
	(0.68)	(0.02)	(0.08)	(0.74)	(0.39)
PERPOP	-0.374	-0.485	-0.691	0.164	0.498
	(5.72)	(3.60)	(1.70)	(0.64)	(3.34)
CONSTANT	-9.93	-7.07	-39.01	1.64	-2.40
R^2	0.96	0.54	0.86	0.93	0.80
F	41.52	2.85	10.43	22.83	7.31
SE	0.05	0.17	0.08	0.09	0.10
DW	2.91	2.57	2.65	2.69	2.55

overall F test, F(12, 7, .05) = 3.57. (2) Acronyms are defined in Appendix B,

APPENDIX B
LIST OF ACRONYMS

AGE	Average age at which each cohort started the doctorate
AIP	American Institute of Physics
BATTELLE1	Average real salary for baccalaureates in chemistry, physics, engineering, and life sciences 10 years post-baccalaureate (obtained from Battelle Columbus Laboratories)
BCARN1ST	Percentage of a cohort with a baccalaureate from an institution classified as "Research I" by 1987 Carnegie Classification of Institutions of Higher Education
BCARN2ND	Percentage of a cohort with a baccalaureate from an institution classified as "Research I or Research II" by 1987 Carnegie Classification of Institutions of Higher Education
BCL	Battelle Columbus Laboratories
BCPCREAL	Average real starting salary for baccalaureates in a particular field (obtained from the College Placement Council)
BLACK	Percentage in each cohort who are black
BSALPROF	Average real starting salary for baccalaureates in a particular field (obtained from the American Institute of Physics for physics and astronomy and from National Survey of Hospital and Medical School Salaries for health sciences)
BSALREAL	Average real starting salary for baccalaureates in a particular field (obtained from the Endicott Report)
BTOP20	Percentage of a cohort with a baccalaureate from an institution with a graduate program ranking in one of the top 20 according to the National Research Council's 1981-82 Assessment of Research-Doctorate Programs in the United States
BTOP40	Percentage of cohort with a baccalaureate from an institution with a graduate program ranking in one of the top 40 according to the National Research Council's 1981-82 Assessment of Research-Doctorate Programs in the United States
CASPAR	Computer Aided Science Policy Analysis and Research System of the National Science Foundation
CPC	College Placement Council

DEFIN	Percentage of the DRF cohort who had definite employment or postdoctorate study plans at the time that the survey was conducted
DEPEND	Average number of dependents of the doctorate recipients in each cohort
DRF	Doctorate Records File
E	Market forces (i.e., SALRAT1, UNEMP4YR, SEEK)
EAM	Earth, Atmospheric and Marine Sciences
ER	Endicott Report produced at the Northwestern University
EXPENDST	Federal expenditures on higher education per full-time equivalent student
F	Family background characteristics (i.e., MARRIED, DEPEND, TEMP, WOMEN)
FACULTY	Number of full-time equivalent faculty members
FORBACC	Percentage of each cohort with a foreign baccalaureate degree
GREQ	Mean quantitative score from the Graduate Records Examination
GREV	Mean verbal score from the Graduate Records Examination
HISP	Percentage in each cohort who are Hispanic
I	Student attributes (i.e., AGE, SAMEFLD, SELECT)
MARRIED	Percentage of each doctorate cohort that are married
NSHMSS	National Survey of Hospital and Medical School Salaries produced at the University of Texas
O	Institutional environment and policies (i.e., FORBACC, BTOP40, BCARN1ST, BCARN2ND, PTOP40, PCARN1ST, PCARN2ND, FACULTY, R&D)
P&A	Physics and astronomy
PCARN1ST	Percentage of a cohort with a doctorate from a "Research I" institution, based on 1987 Carnegie Classification of Institutions of Higher Education
PCARN2ND	Percentage of a cohort with a doctorate from a "Research I or Research II" institution, based on 1987 Carnegie Classification of Institutions of Higher Education
PERPOP	Ratio of number of doctorates to the U.S. population 25-34 years of age having 16 or more years of education
PTOP20	Percentage of a cohort with a doctorate from an institution with a graduate program ranking in one of the top 20 according to the National Research Council's 1981-82 Assessment of Research-Doctorate Programs in the United States
PTOP40	Percentage of a cohort with a doctorate from an institution with a graduate program ranking in one of the top 40 according to the National Research Council's 1981-82 Assessment of Research-Doctorate Programs in the United States

R&D	Research and development; ratio of real dollar value of government expenditures on university R&D to the number of science and engineering doctorate recipients
RTD	Registered time to the doctorate (i.e., the length of time that a student is actually registered in graduate school)
SALRAT1	Ratio of SDRSAL10 to SDRSAL
SALRATIO	Ratio of doctorate salary to baccalaureate salary in a particular field
SAMEFLD	Percentage of each cohort with a baccalaureate degree in the same field as the doctorate
SDR	Survey of Doctorate Recipients
SDRSAL	Average real salary of recent doctorates in a particular field (based on SDR data)
SDRSAL10	Average real doctorate salary for doctorates 10 years after receipt of the degree (based on SDR data)
SEEK	Percentage of those in a DRF cohort seeking employment or postdoctorate study
SELECT	Percentage of each cohort from selective undergraduate colleges and universities
SUPFED	Percentage of each cohort with federal support
SUPOWN	Percentage of each cohort reporting own earnings as primary source of support
SUPPRIV	Percentage of each cohort with private foundation support
SUPRA	Percentage of each cohort with research assistantship
SUPTA	Percentage of each cohort with a teaching assistantship
TEMP	Percentage of the total doctorates who hold temporary visas
TLFA	Tuition and financial aid (i.e., TUITION, SUPFED, SUPPRIV, SUPTA, SUPRA)
TNEU	Time not enrolled in the university after beginning graduate studies
TPGE	Time after receiving the baccalaureate but prior to graduate entrance
TTD	Total time to doctorate (i.e., the time lapse from the year of receiving the baccalaureate until the doctorate is completed)
TUITION	Real average in-state tuition and fees paid by the cohort
UNEMP	Overall unemployment rate for the U.S. labor force (obtained from the Bureau of Labor Statistics)
UNEMP4YR	Unemployment rate for persons with four or more years of college (obtained from the Bureau of Labor Statistics)
WOMEN	Percentage in each cohort who are female